Voyager

THE WORLD FLIGHT

THE OFFICIAL LOG, FLIGHT ANALYSIS AND NARRATIVE EXPLANATION
OF THE RECORD AROUND THE WORLD FLIGHT OF THE VOYAGER AIRCRAFT

THE OFFICIAL LOG AND FLIGHT ANALYSIS IN ALL THE OFFICIAL RECORDS AND ARCHIVES

BY

JACK NORRIS

TECHNICAL DIRECTOR

VOYAGER MISSION CONTROL

Northridge, California 1988

Text Editor: Milly Norris

Library of Congress Catalog Card Number: 88-90609
ISBN 0-9620239-0-6

Manufactured in the United States of America
Second Printing January 1989

VOYAGER, THE WORLD FLIGHT

THE EXPLANATION, COMPOSITE LOG AND ANALYSIS OF
THE WORLD FLIGHT OF THE VOYAGER

The milestone unrefueled World Flight of the Voyager is arguably the
ultimate atmospheric flight, the greatest airplane flight in the history
of man's continuing conquest of flight, "the last great first" of the
original airplane.

The accomplishment of World Flight was a magnificent technical achievement,
but it was also a great human people story, and ultimately a real life
adventure thriller. It is such a great story in total that it needs to
be told in understandable form, in a way that all those interested in flight
can understand and appreciate just what did happen, what was done and how
it was done.

The story herein started as the official analysis of the flight by the
Technical Director of Voyager Mission Control as a companion piece to back
up the official NAA/FAI record about the five designated official checkpoints. The five official checkpoints naturally produce shorter direct
mileage than the actual total, and the two ways of looking at the flight
become natural complementary methods, each valid for their own purposes,
not conflicting in any way. This official analysis was made to be placed
in all the official museums and archives as the complete and accurate
historic record, since there is and can be no other record of the flight
that is so complete and so accurately checked and cross checked with all
the available data and experienced understanding of the true facts of the
flight.

To make it more understandable to those interested in flight and to laymen,
the original technical analysis of the flight was fleshed out with descrip-
tive narratives that explain what was really done and how it was done, so
that everyone interested can better understand and better appreciate the
great technical and human achievement involved.

The original official technical analysis of the flight was left in its
original form, even with the original hand drawn flight analysis graphs
in the rear of the book, so that the book can be read at several levels.
The layman can get a very good grasp of what was done and how it was done,
by simply reading the early narrative portion of the book and looking at
the composite log itself. Those interested in the specific technical
flight details can go deeper and deeper into exactly what happened by
simply reading farther and looking closer. Those who simply wish to
understand can simply skip all the exact analysis, numbers, and technical
explanation and get the degree of understanding they desire.

The format produces some repetition of key points, but if you will excuse
that minor flaw of form, you will see that it makes it possible for
readers at all levels to correctly understand the key points and conclusions. For historical purposes the book was purposely not reformated
but was published in finished draft form.

A YEAR LATER, DECEMBER 14, 1987

The fundamental objective of the Voyager Project was to audaciously set out to double the flight distance record and accomplish unrefueled World Flight, an obvious Major Historic Milestone of Flight. The Voyager did the job, and it ended up right along with the Wright Flyer, the Spirit of St. Louis, the X1 and all the spacecraft, the true milestones of flight, properly honored.

The first copies of Dick and Jeana's book were available to us on this first anniversary. As might be expected, there are places where the book, written from the plane's log, in parallel with this analysis has different numbers and conclusions than those found herein. This report has the numbers that are cross checked and are the ones that should be taken as correct.

By quickly reading the book, we, in turn, were able to correct ourselves in finding that it was Dick, not Jeana, who physically switched the failed fuel pump off Mexico, a task we had attributed to Jeana in the initial copies.

The book, in turn, makes abundantly clear that Jeana, a wonderfully strong, determined, intelligent lady held the mission together several times when Dick, an equally impressive and deserving modern day hero, predictably, got right to the end of his rope. They were an amazing team.

————————

This was the second time around for me. I had been deeply involved in the space program, having made a majority of the control components in the Mercury, Gemini and Apollo small rocket orbit and maneuvering systems. I know what these flights take, I had seen my equipment go into the National Flight Museum the day it opened. With twenty years perspective, I could contribute best by seeing and plugging holes that others wouldn't see and anticipating what would happen before it happened.

The Mission Control people were first brought together as a training exercise on the Pacific Coast Closed Circuit Record Flight in July, 1986. Naturally people were not organized then, the task was to learn what was required, what had to be done. Returning in mid flight, Burt Rutan, the Voyager designer, was frustrated because no one was working on performance. Burt is extremely smart and typical of brilliant people, hoped for more than people were going to deliver. He didn't realize that he hadn't told anyone what was required or how to do it, assuming that they would understand and know.

As a Management Consultant these days, I'm used to seeing that kind of gap and looking, I could see that there was no one whose orientation and background would equip them to take on the complex performance analysis job. Despite having gotten past number crunching many years before, I dove into what was to become a difficult, even dangerous job. Ultimately, however it was a great satisfaction, a thorough education on aerodynamics, the inner secrets of flight, and a lot of fun, because it was, as you will see, a real challenge. It was in fact a historic opportunity.

———————

The long delayed December launch decision was made so quickly that many of the key Mission Control people were not even there at launch. Going full bore to launch with no last minute retraining and agreement on ground rules, with completely new routing, new speed and altitude profiles just barely worked out in time, there were all kinds of opportunities for confusion and mistakes, that were professionally handled on the run. We commented at the time that it could not have been more intense if it were the bunker running World War III.

All of the key people on the ground as well as in the plane were under terrific stress, faced with great difficulty, with great responsibility, many with little sleep on this nine day adventure.

We knew from our planning studies that this was the wrong time of the year and that it was going to be pure hell. It was our job to know when others didn't that there is a huge difference between a winter and a summer flight, and that we would not find a good weather time in December. We tried lightly to talk Dick and Jeana out of the go decision, but to not make the launch decision even more difficult than it was for them, since they felt they must go to keep the team together, we got out of the way, after giving them a shot at the truth. We then stayed out of the predictably difficult communication loop, let the pro's there do their job, while we did ours. Everyone came through, but by no means was it easy on anyone.

———————

The final game of course was to not run out of gas and Dick put great effort into an engine fuel flow system that would read out to the needed $\frac{1}{2}\%$ accuracy, a difficult technical challenge. On the World Flight, however, weather pummeled Voyager enough that it biased the data between reports, making it physically impossible to calculate a sufficiently accurate "fuel remaining" answer -- no matter how you did it! We were sufficiently prepared, however, that we spotted the problem and by comparing the misleading 9 day running calculation with our incisive planning, we kept under control despite what you will see was a ridiculously complex challenge.

Dick had a love-hate relation with the Voyager. Because of its extreme load and stability problems, he was intelligently afraid of it and was so locked into the herculean task of getting the plane ready and doing the flight, he wasn't thinking before launch just how much trouble fuel analysis could be if things didn't go as he intended. The book shows that in mid flight, exhausted, he would be fully frustrated trying to get the performance data through the very difficult communications. Without it of course, we simply could not do the basic Mission Control job of evaluating and guiding the flight. Also, having built and flown the world's most fantastic airplane, it would be a great mistake to not fully understand and be able to make an accurate account of what it really did on its epic flight. This was not a NASA funded program with a building full of engineers and the difficult facts were simply going to get lost if we failed.

As you will see, understanding what was really happening to the fuel situation aboard the aircraft, despite false information from both of the fuel data sources, became one of the major challenges of the flight.

You will see that despite the mind boggling false data and calculations, we stayed under control and our bacon was saved because, as expected, the intrepid Voyager came very close to meeting the planned performance curve. Despite all the weather challenges, our Flight Plan worked, and we thus had a well thought out sound reference to steady our judgment! Anticipating trouble, we were prepared and lived through a wild nine days!

Even the other people in Mission Control didn't understand the problems of the flight analysis, because that terribly challenging hectic 9 days was not the time to try to teach people the hidden complexities of a Voyager Flight. Furthermore, since the Voyager was truly a technical triumph in every way, we didn't want some controversy seeking reporter to use it to try to inject controversy and tarnish the image of the Voyager. In the year of the Challenger, it was time to have a human triumph reported as the triumph that it genuinely was.

Dick, Jeana and Burt, despite all kinds of doubters brought off one of the genuine major accomplishments in the history of flight and are genuinely deserving of every reward and praise that they have received. Naturally everything wasn't perfect, but in the proper overview perspective any short-comings are vastly unimportant. It's worth noting that when a major triumph like this is achieved there is almost always a significant underestimation of the time required and the number and degree of difficulties that must be overcome. It is absolutely standard on a project like this to take much more time than was anticipated and to find in hindsight that a herculean effort was required. Achieving the objectives despite all the challenges is the final validation of the people.

The bottom line story on the Voyager was that it was a magnificent technical triumph by people that genuinely had all the Right Stuff. Not the Hollywood kind, the real Right Stuff: genuine intelligence, creativity, perseverance, guts, unbelievable stamina. The Astronauts and the Chuck Yeager's are super people, but these people also had to design and build their own plane, keep the bills paid for five years and fly a mission that would have broken the back of an iron man. The more you learn herein, the more you'll understand that Voyager was a real triumph of the human race. It beats Jules Verne hands down because this one was real.

GLOBAL CIRCULATION •Tropical Low Altitude Easterly Winds
•Intertropical Convergence Zone

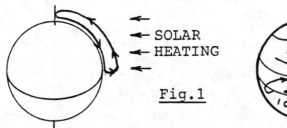

Equatorial Air
Rotating 1000 MPH
Flowing Poleward
Swings Eastward
Become Westerlies.
Some Cools, Sinks,
Returns To Equator
As Low Altitude
Tropical Easterlies

SOLAR HEATING

Fig.1

NON ROTATING WORLD   ROTATING WORLD

Converging, Rising Equatorial Air Masses Cause The ITCZ

GLOBAL CIRCULATION

THE FLIGHT OF THE VOYAGER

Understanding global flow is the first key to understanding where and when you should make a world flight and why the flight of the Voyager became an adventure thriller.

The sun heating the Equatorial Latitudes drives a global circulation, where the heated Equatorial air rises and begins a circulation toward the cold, unheated Poles. The Equatorial air, however, is rotating at over 1000 miles per hour, since the earth's nominal 25,000 mile circumference rotates once each 24 hours. Since the air would slow to zero rotation if it circulated straight to the Pole, the circulation bends sharply Eastward becoming a high altitude West wind as it proceeds Poleward. By the Horse Latitudes, 30° North and South, some of the air has cooled and sinks and becomes a low altitude Northeasterly recirculation in the Northern Hemisphere (Southeastern in the Southern Hemisphere). They become the Northeasterly and Southeasterly Tropical Trade Winds that blow steadily in the Tropics, ultimately almost straight Westerly from the East, as the two great masses of air converge at the earth's thermal equator. See Fig. 1.

Unfortunately, this Intertropical Convergence Zone (ITC) forms giant cells of very unfriendly weather for a heavily loaded aircraft that must be avoided at the thermal equator, but the adjacent Tropical Trade Winds flowing steadily over long stretches of ocean from East to West provide the ideal way to fly a world flight. The Northeasterly winds available at low altitudes out of Los Angeles in our summer are the equally ideal way to get to the Tropics from Los Angeles.

The most favorable place and time to fly a world flight is over water in the band of maximum low altitude Easterly Trade Winds, averaging 10° to 12° South of the equator, at an altitude of 3,000 to 7,000 ft., in the June to August full moon periods. On the winter side of the equator, in the benign dry season, the ITC, the broken band of bad weather that is the earth's thermal equator, has moved North for the Northern Hemisphere Summer. Ideal world flight conditions result from all this, optimum, essentially continual tailwinds, benign tropical weather over broad undisturbed ocean stretches.

It is not possible to find a Northern Hemisphere equivalent optimum in winter. The earth is not symmetrical. There is much more land North, which produces much more heating. This extra Northern heating holds the ITC back from moving South enough to be out of the way of a Northern Hemisphere flight.

A Northern Hemisphere flight thus is forced to fly along the turbulent weather laden Northern edge of the ITC. In attempting to fly in the best wind band mostly over water, dodging the typhoons of the South Western Pacific, the ITC forces the flight up against the unfriendly South Asia border of Vietnam, across Southern Thailand, up against Sri Lanka and the Southern tip of India and toward the greatly wider, more turbulent Northern half of Africa, over the dense equatorial jungle. A crossing of the thin waist of Africa just above or below the equator, between unfriendly, unstable countries, where flights have been shot down, or could be harassed, would be the best of the worst, but is only marginally available.

South America is bypassed by leaving for the Tropics from California and returning to California from the tropical Atlantic off the Eastern tip of Brazil. Africa is the key problem.

When the flight was ready to go in December 1986, the launch decision was made in the full knowledge that the flight would be made the hard way, not the easy way, with weather a factor along the entire tropical route.

Whereas many were concerned that the crew would not be able to make it through the arduous 9 day flight, Voyager Mission Control knew that there was no doubt whatsoever that the tough, trained crew would be able to fly the mission in the optimum June - August, full moon periods when any weather could also be seen at night.

When Dick and Jeana decided to launch in the much more difficult December period, it was clear that they would be forced to cope with the predictable mechanical problems and flight emergencies while also being pummelled and fatigued by continual weather problems along the majority of the route.

The key fact to understand is that the flight was not made at the right time, via the right route, but rather it was made despite the fact that it was a most difficult time and a most difficult route. That added an adventure thriller on top of a classical technical and human achievement.

It was absolutely predictable that the crew would be not just short of sleep, but bone tired when they encountered trouble, that crew fatigue would be a major factor rather than a non factor, that real risk was being introduced into a flight, that through hard work and good management, was otherwise becoming more and more of a sure thing, subject only to the irreducible risk of mechanical failure and reliability problems.

The flight and the crew did encounter every sort of weather problem, mechanical problem and flight emergency that could be predicted. It became an adventure, a test of human will and stamina worthy of the best of Jules Verne or a George Lucas movie. The crew's stamina and performance was remarkable.

The World Flight of the Voyager was a piece of world history, it was a major milestone of flight, equivalent in every way to that of the Wrights, Lindbergh, Chuck Yeager and the Space Triumphs. It was a magnificent technical achievement, a triumph of composite structures and incisive, imaginative, creative aerodynamic design. It actually doubled the world flight record in a high technology age where Olympics and world records are normally raised by a small percentage -- and it was designed and built by the people who did it.

It was in total a splendid example of the will and achievement of the human race, a human "only in America" people story, where a few free Americans started in pursuit of a dream and putting aside every obstacle, did mission impossible on a grass roots basis unaided by government, far surpassing what the best of corporations or organizations had attempted or even dared to dream.

THE ACCOMPLISHMENT OF MISSION IMPOSSIBLE

It is much more than twice as hard to fly twice as far unrefueled. On first appraisal, it would appear to be impossible.

The B52, which held the previous distance record of 12,532.28 miles, half way around the world, was the essentially optimum distance speed and load carrying product of the full resources of the fully funded Aircraft Industry of the United States of America. When it took off, it would have been at its maximum load, maximum takeoff capacity.

If twice the fuel were somehow loaded, it would not take off, it would not fly, it would not even have been safe to taxi. Most significantly, if it had been able to fly, it would not have flown twice as far, because optimum airplanes fly on a lift/drag ratio, and if one weighs twice as much with fuel and the structure to carry it, its lift and drag would be twice as much, and it would initially take twice the fuel energy to fly each mile.

Long distance flight becomes a compounding problem of dimishing returns, and flying twice as far is indeed impossible unless an approach can be devised that is indeed twice as energy efficient.

Doubling the energy efficiency, that is using half the energy each mile, again initially appears impossible in this already very efficient well engineered world that we live in. Engine and propeller efficiencies must be raised to near optimum levels, but it is not possible to double their efficiency. A propeller at 80% efficiency can not be doubled to more than 100%.

The airplane itself must be essentially twice as efficient. Its drag must be greatly reduced. It would be necessary to essentially double its L/D ratio, but since that will not be possible, its structural efficiency must be far greater, so that it can carry a greatly larger fuel load without a corresponding structural weight penalty. Double range and margin results.

You will see herein that the final airplane that resulted is first a structural wonder, a product of the most modern carbon fiber/epoxy composite construction that is magnificently strong while being magnificently light, far beyond the capability of a metal airplane.

The airplane will be a wonder of aerodynamic efficiency designed to achieve very low drag with laminar flow, and a very long structurally difficult wing that greatly lowers lift drag, so the huge fuel load can be carried without a huge drag due to lift.

The huge initial, but ever decreasing load requires the staged power of two engines, but also very high efficiency over broad power ranges. The main rear engine was the new Teledyne Continental liquid cooled engine that runs over the necessary broad power range at efficiencies that average to performance levels that were once held to be out of reach optimums. The front engine is shut down as the huge load burns off.

With optimally efficient rugged TRW Hartzell propellers, built to optimized computer designs on a crash basis to save the program, there is optimum conversion of fuel energy to thrust and horsepower to overcome minimum aircraft drag. The flight efficiencies achieved made mission impossible real, indeed with the necessary margin to overcome the practical problems of battering by world weather, detours, off speed flight, even bug impacts inhibiting laminar flow locally.

It was the creativity of Burt Rutan that produced the unique configuration of the Voyager and extremely high structural and aerodynamic efficiency that made World Flight possible. The long structurally difficult wing is only possible if the fuel load is spread over the wing to reduce bending. Since the volume of the wing would be insufficient to carry the fuel, large booms are placed outboard to carry the fuel and spread the load. Huge booms would have twisted the wing to failure or instability, but by tying them to a stiff forward cunard wing, that itself carries a full measure of fuel and lifts up, whereas a normal tail pushes down, a self reinforcing configuration of unique structural and aerodynamic efficiency was created.

Long range flight is done by controlling speed vs weight to hold a constant optimum angle of attack to achieve and maintain the optimum flight efficiency that the airplane can deliver.

Technical Marvels, Impossible Missions are accomplished by exceptional human beings. Dick Rutan and Jeana Yeager built Voyager with their own hands, sweat, brain power and perseverance without adequate or available funding over a five year effort. They were helped by Bruce Evans, the first man in a growing cadre of volunteers who shared and supported and helped create the dream of World Flight. Burt Rutan was brilliantly creative in understanding and conceiving the solution to Mission Impossible, but he was also the accurate practical designer that almost single-handedly produced a flaw free design that did not fail and that at his direction, even had the substantial range margin that he foresaw as necessary to allow for practical pitfalls. Seldom is a creator also that accurate and practical.

This account of the World Flight of the Voyager started as and is the Official Flight Analysis, Composite Log and Performance Report of the World Flight by the Technical Director of Voyager Mission Control. If you simply read the report in sequence, you will see that you will be provided with all the necessary information to understand the plane and the flight, the sweat in the cockpit and the sweat in Mission Control on the World Flight.

When you finish, you will know and better appreciate that you can indeed be proud to live in a country like America, which produces people with such outstanding initiative, creativity, intelligence and perseverance. When individuals at their own initiative and with their own hands can produce technical marvels that exceed the accomplishments of the largest corporations and government, we indeed have something that is very important to preserve.

UNDERSTANDING THIS PRESENTATION

In this booklet, narratives that lead you into correctly understanding the World Flight of the Voyager are presented along with the refined composite log of the World Flight, the flight analysis and comparisons to the planned flight, and to a theoretically perfect flight, where the Voyager would follow the performance curves perfectly.

A refined log is needed to combine and correct all the air to ground data and the data on the aircraft flight log, to put together all the facts in one place for the correct final grasp. The Airplane Log is used as the basis of the refined log and analysis, but since it has gaps, the Air to Ground Log is used to fill the gaps and produce a complete accurate log. Minor corrections, for example, making the time of position reports more accurately apply to the position in those cases where the radio log time was inaccurately applied to the position, make the log numbers valid for calculation and overall analysis. The log data was used for altitude data, rather than the barograph, but to better understand the climbs and to be able to more accurately calculate climb fuel, barograph insight was used there. The refined log corrects the fuel burn error of the raw data log!

The bottom line numbers on the log give the facts on the World Flight which are very interesting and of historical importance -- but what really happened on airplane performance and how long range flight really works is best seen by comparison to the planned flight and separately to a theoretically perfect flight using the actual distance and fuel load numbers.

A narrative comparison of the actual flight to the planning flight along with interesting insights on problems provides an interesting story and facts that anyone can understand, and a feel for the flight.

As the report goes on and into the Engineering Appendix, the specific engineering comparisons are made for the people who really want to dig and know the specifics and grasp the real inside story. That gets even more interesting but is not easy and must be earned by tackling a much more complex investigation.

You will see that the format here is to add pages that provide you with the insights and the facts that lead you into a proper understanding of this historic flight.

When you come to an early section that looks too technical, too complex, read it anyway. You will find that it was written with understanding by a non technical person in mind. Even if you miss or skip some of the content or the numbers, you will find that you understand most of it. If toward the end of the report, it becomes too technical, too deep for you and you stop, you will find that you already understand how the World Flight of the Voyager was accomplished. In sections where there are too many numbers (because this is the Official Flight Analysis) just skip numerical minutia and grasp the salient facts and understanding.

WHAT ARE WE REALLY DEALING WITH HERE?

The Flight of the Voyager is really an incredibly complex, ever changing, interdependent calculus problem done by thousands of sequential arithmetic calculations done either by hand calculator or computer. Weight and therefore proper speed and rate of fuel consumption is ever changing. Winds vary and as altitude is changed true airspeed changes along with power requirements and rate of fuel consumption, which all tie back into rate of weight change, speed and time. The time of arrival and fuel remaining becomes a morass of numbers, an impossible task if you're not prepared.

If you naively sit down to calculate a Voyager Flight using the data directly from the plane, you get wrong answers. The nine day Voyager Flight is not a normal airplane flight and the usual simple arithmetic does not always work. If you use 100 pounds of fuel climbing in mid flight, as we did, you assume it costs you 100 pounds but it does not. The obscure mathematics of the Voyager causes it to only cost 73 pounds ultimately! Even the other experienced pilots in Mission Control didn't understand what was happening during the World Flight and those hectic nine days was not the time to try explaining where all the obscure analytical traps were.

Coping with this incredibly complex problem when the World Flight was on was a real piece of work. Voyager Performance had done the necessary homework, however, and understood not just the problems and the range of answers and traps, but the obscure fact that the numbers available during the flight would not yield correct answers due to a built in optimistic reporting bias. Quite simply, the inflight performance reports were transmitted when things tended to be under control, but between reports weather challenges and problems would upset the reported numbers by just enough that a small error, but one too big to ignore, worked its way into the data available for calculating the flight. We saw that in time, avoided bad answers.

By having the planning numbers and the months of necessary experience, we knew with certainty that the flight would make it and with what potential margins as long as we did not run into a mechanical failure or an insurmountable weather block. We also had an inflight cross check on the calculated numbers and knew the very important fact that if the inflight numbers could not be trusted that the final fuel checks as the tanks ran dry would provide the necessary end of flight checks. They alone provided a dependable and accurate end of flight check of the remaining fuel.

With everyone working with very little sleep due to the continual challenges of the flight, as the narrative and the facts included herein show, we did very well indeed. We even managed to know at flight end that the wind had been quite nominal and that our early arrival was due to higher flight, not extra winds -- and that our headwinds were off Mexico, not over Africa.

The story is all in here. First in simple understandable summary and narrative form and then the specific engineering numbers for the incisive.

Even if you quit before the more difficult engineering sections in the rear, you will end up understanding what happened here very well.

A QUICK LOOK AT A FEW FINAL CONCLUSIONS

There are two numbers, 286 pounds of fuel, the cost of lost performance (most of it voluntary) and 243 pounds of fuel, the penalty for flying heavier than plan, that become the two keystone check numbers that finally tie together and check the thousands of numbers that describe the World Flight of the Voyager. More on that shortly.

This booklet is purposely written so that everyone, anyone who is interested can understand the fuel usage and performance numbers on the World Flight of the Voyager through the refined log and through a simple, understandable narrative that compares the World Flight with the plan without grappling with the mind boggling complexity that is really there.

For the few who will go all the way into and through the Appendix where the engineering analysis is hidden, you will find a numerical Sherlock Holmes mystery that both dissects the World Flight numbers as it was -- and specifically compares it to the theoretical performance curves and the planned flight and shows both the fantastic comparisons that are possible and the fundamental differences.

> Everyone will be able to see that the Voyager was a miniscule 3 gallons, 18 pounds of fuel off the plan at the Caribbean decision point before it was voluntarily decided to come home rich and safe. The Voyager target was to return with 400 pounds of fuel, a 2400 Nautical mile transcontinental reserve range. Before the decision to come home rich, the Voyager numbers were heading for 405 pounds, less than 1 gallon off. Naturally there was luck in such an incredibly close outcome. It is equally true, however, that having such good luck was only possible because we knew exactly what we were doing. The entire Voyager effort was a first class piece of work by the entire team.

There's a fantastic numbers story here that matches the world class story that the Voyager was. It is expected that only a few, the numerically sophisticated and those interested will go or be able to go all the way through the complete analysis presented here, but those that do will be rewarded by really understanding the true intricacies of the Voyager Flight that lies behind the simple story presented for everyone.

286 pounds of fuel, 1469*Nautical miles is the exact measure of range lost due to below curve performance on the weather challenged World Flight -- but 190 pounds was voluntary, a purposeful decision to come home rich and safe, so that the real underachievement was less than 100 pounds of fuel (96) in the 7011.5 pounds of fuel that the Voyager carried, the score of a fantastic airplane and crew.

Burt Rutan's calculations concluded that the wingtips damaged on takeoff cost 115 pounds of fuel. The conclusion above that the plane actually flew 96 pounds below the planning curve indicates that the core airplane flew within 19 pounds of fuel (3 gal) of the planning curve. The Voyager in test consistently flew 5% or more above Burt's original minimum Specific Range Curve as shown in the Appendix. In hindsight, the conservative decision to use the minimum curve for planning the more challenged World Flight could hardly have been better.

*1691 Statute Miles

As an airplane gets heavier, it takes much more power and fuel to fly it.
If you put in twice as much fuel, it does <u>not</u> go twice as far! <u>Heavy, it</u>
<u>uses fuel much faster!</u> Flying twice as far as any other airplane, flying
all the way around the world is a huge problem.

Fuel is both the load and the source of energy to lift and carry the load.
If you add a pound of fuel most of it is used up carrying itself around
the world, so there is very little left at the end to fly farther! <u>That</u>
is the fundamental problem of very long range flight! The plane finally
becomes so heavy it won't take off! If you do get it off, it flies little
extra distance, because the fuel uses itself up carrying itself around!
Only a magnificently efficient aircraft can do what the Voyager did!
The structure must be unbelievably light and unbelievably strong, so that
the fuel load can be huge! - - - And there is no room for error!

THAT IS THE PROBLEM THAT BURT AND DICK AND JEANA FACED - - along with a
WORLD OF WEATHER - - NINE DAYS OF ARDUOUS DUTY - - WITH LITTLE SLEEP!

It cost 243 pounds of fuel to fly the Voyager 394.5 pounds heavier than
planned, 83 pounds of extra cabin weight and 311.5 pounds of extra fuel.
Thus the 311.5 pounds of fuel only provided 68 pounds of extra fuel at
the end -- which gives you an incisive grasp of the importance of the
fantastically lightweight structure of the Voyager -- which provides
super range at the end, light vs the low payoff heavy visible in the
68/311.5 pound ratio. (68#, 11.7 Gal. is 2.4 to 6 Hr., 190 to 380 Nm.)

Those who go all the way through to an incisive grasp will find these
two numbers, 286 pounds and 243, tie the final technical analysis together
and check the final calculations.

Though no attempt was made to get exact agreement on the final check
numbers, the final check numbers did in general come out to less than
one pound, a fitting cap for a fantastic flight, plane and crew, a
fitting finale to the major milestone of flight that the Voyager was.

While the Voyager was a fantastic, World Class technical achievement, it
should be clear to you already that in the last analysis, Voyager was a
World Class people story. Burt and Dick and Jeana are the super achievers
who did it. Magically drawn by the mystical, imaginative, major milestone
goal of world flight, a whole cadre of right people came together to support
it. Thousands of VIP's gave $100 each to fund it. Chuck Richey, Burt's
Chief Engineer, designed an amazing 24 pound per strut landing gear. John
Roncz, an analytical genius, designed special airfoils and special propellers
for Burt. Bruce Evans, joined by Ferg Fay, Glenn Maben and others, became
the perfect crew. Chuck and Joan Richey, Mike and Sally Melvill, and a
host of others were friends and early helpers. Larry Caskey and Len Snellman
were the nucleus around which a genuinely professional Mission Control formed.

FORWARD

It was clear to those who understood that the Voyager was going to be a true
piece of history, a major milestone of flight and a classic story of human
vision, creativity, determination, perseverance, the stuff that has made
America what it is and that in turn would only happen in the freedom of
America where individuals can and do what only the free and the bright can do.

As such, the Voyager deserved a first class log and analysis befitting its
historical importance both for the record and to better understand what
really happened.

With 6,510 "slots" in its log, 217 check points, 30 possible data items for
each point, it was far too extensive to calculate and refine multiple times
by hand and too varied to program for computer analysis. A special compliment
and note of thanks is due Brian Hobbs of the Flight Simulation and Analysis
Branch at Edwards Air Force Base for the tortuously complex exception based
program that made possible the hundreds of thousands of calculations that
permitted the refinement and print-out of both the raw data version of the
log and the final, refined, analyzed log.

There are three objectives herein.

1. To provide the specific facts, the log and the supporting facts
   and analysis of The World Flight.

2. To provide an understandable interesting narrative that makes clear
   to all who are interested what really happened, what analysis of
   the extensive data shows -- in the full understanding that the
   conclusions and understanding, not the analysis of data is
   what we humans are interested in.

3. For those who do seek to look deeper, a precise engineering analysis
   and some of the backup data are included in the rear of the report.

It was as you shall see a magnificent technical accomplishment and it became
a classic human adventure thriller challenged all the way by weather --
and yet, as you will see, it flew so close to plan, it's amazing.

ABOUT THE PRINTING OF THIS BOOK

Rather than retype or typeset this book for perfect professional formatting,
it was purposely left in its original form as initially completed so that
you could have an original, an even more valuable record of a genuine piece
of flight history.

In the second printing, the formal name of our National Museum has been
deleted, since it is a registered trademark. A few new insights, words
and paragraphs have been added where space permitted to be more clear.
Both first and second printing are filed as the historic record, just as
you have it.

THE PERFORMANCE GROUP

Now we can tell our secret. We in the Performance Group had the best jobs
in the program after Dick, Jeana, Burt and Bruce, the people who really did it.

Coming in to seriously help the super work of the genuine originals, we took
on what in most cases would be the dull green eyeshade job of number crunching,
often given to smart recluses, but that is not what happened here at all.

The Voyager performance was the solution, but finding out the real definition
of the flight problem and the Voyager's ability to do the task was second
only in interest, intellectual challenge and fun to the magnificent technical
job that Burt did, or strapping on and going as Dick and Jeana could and did.

Most of the effort of the Performance Group was not dull numbers. After
everyone else did the five year's work, we got to figure out where it was
really going, what the wind and weather was really like, whether the plane
would really make it and with what margin, a real total understanding of
the flight before it happened -- AND HOW TO OPTIMIZE IT!

With the DWIPS world-wide satellite weather data at our elbow for half a year,
with the best of weather insight ever assembled, from Len Snellman, from new
friends like Keith Gordon of QANTAS, who provided the first clue to the
optimum aerial railroad track that exists, the chance to sift and sort the
route, weather, winds and the plane's capability added up to one of the classic
best jobs in the history of flight for someone interested in flying.

To keep people thinking and flexible, Dick properly kept the flight represented
as a free form flight to go wherever the conditions were best. The fact is
however, that if you really study it, the place where the weather and the
winds and the geography are best, both North and South are so clear there are
world-wide slalom gates that you fly through and you might as well be on an
aerial railroad track except for minor chance weather diversion. We flew the plan.

When the plane reached the Caribbean decision point, it was 3 gallons, 18 lbs.
of fuel ahead of the final plan given to Dick the afternoon before the flight.
Before the rich return home was instituted, it was headed precisely for the
400 lb. reserve (full tip tanks, plus some in the feed tank) that Dick had
directed the summer before.

The names of the Voyager Performance Group are proudly included on the log
of the Voyager because, as you will see, the homework had been done and done
right and the plan worked.

Jim Richmond	Route Optimization Analysis
Milly Norris	Lockheed Data Plan Flight Analysis
Terri Smith	Lockheed Data Plan Flight Analysis
John Burns	Analyst, World Flight
Eric Knutson	Analyst, World Flight
Jim Johnson	Analyst, World Flight
Kelly Sandfer	Flight Log Preparation
Brian Hobbs	Assistant Technical Director

Jack Norris

VOYAGER WORLD FLIGHT: EDWARDS AIR FORCE BASE CALIFORNIA[2]

Start Takeoff,Runway 4   07:59:38 Sunday   Dec. 14, 1986 (Temp 30°F,Wind Calm)

Liftoff   08:01:44 (2:06) 14,200 Ft (87) 89 Knots

Landing (Compass Rose)   08:05:28 Tuesday Dec. 23, 1986 (Temp 44°F,Wind Calm)

Elapsed Time   9 Days 3 Min 44 Sec (216:03:44)

                                                   (21,712.818 N. Mi)
Official NAA/FAI Distance   24,986.727 Statute Miles around five verified points*
                     +1371.9
Check Point Distance   26,358.6   Statute Miles (22,905.0 Naut Mi)
                     +1455.5
Great Circle Distance   24,903.095 Statute Miles (21,640.19 Naut Mi) NAA Formula

Minimum World Flight Dist. 22,859.79   Statute Miles (19,864.60 Naut Mi)
     (Tropic of Cancer)

	TAKEOFF			LANDING	
Gross Takeoff Weight	9,694.5	Pounds	Calculated Landing Wt	2,699.1	Pounds
Fuel Weight	7,011.5	Pounds	Fuel Wt Landing	106.14**	Pounds
Operational Weight	2,683	Pounds		2,593	Pounds
Aircraft	2,250	Pounds	Aircraft	2,250	Pounds
Provisions	130	Pounds	Provisions	40	Pounds
Crew	303	Pounds	Crew (W/O Wt Loss) 303		Pounds
Structural Weight	939	Pounds			

Average Altitude   9,063 Pressure Alt. 10,540.9 Density Alt. (unweighted)

True Air Speed/Air Miles 112.221 MPH Avg (97.517 Knots) 24,246.7 Air Mi   1.087 f
Ground Speed /Gr. Miles 121.995 MPH Avg (106.011 Knots) 26,358.6 Gr   Mi
Average Tailwind/Wind Mi.   9.774 MPH Avg (8.493 Knots)   2,111.9 Wind Mi

Fuel Used (at 5.8#/Gal)   6,796.4# 1171.8 Gal 22.494 MPG 5.423 Gal/Hr 3.10 N.Mi/#

**190# of the targeted 400# fuel reserve was used for a faster, rich two engine return.
  109# of fuel was lost due to a fuel cap leak on the left tip tank.
  With 106.14# left, the reserve would have otherwise been 405.14# vs the 400# target!!!!
  A 400# targeted reserve at 6 N. Mi/# would be 2400 Naut. Mi., L.A. to N.Y.+!

*Flight Turn Points

EAFB	HAWAII	HATYAI,THAILAND	SUMBURU,KENYA	COSTA RICA	EAFB
N 34°54'18"	N 18°37'	N 06°55'42"	N 00°32'0"	N 10°27'	N 34°54'18"
W117°53' 0"	W158°24'	E100°23'42"	E 37°32'0"	W 85°30'	W117°53' 0"
08:01:44PST	(1248Z)	1646Z	0900Z	0736Z	08:05:28PST
14Dec86	15Dec86	17Dec86	19Dec86	22Dec86	23Dec86
1601:44Z					1605:28Z

Stat Mi 2720.559 + 6799.390 + 4358.189 + 8459.850 + 2648.739 = 24986.727

                        Jack Norris
                        Technical Director, Voyager Mission Control
                        June 20, 1987

HOW MUCH HARDER WAS IT TO DOUBLE THE FLIGHT DISTANCE RECORD?

TO CARRY AROUND ONE EXTRA POUND OF FUEL, HOW MUCH MUST YOU ADD AT THE START?

The World Flight of the Voyager took off at 9694.5#, burned 6796.36# of fuel, landed at 2898.14# before accounting for the 109# of fuel lost and the 90# of provisions used (2699.1#)*. Thus 2699.1#, 27.84% of the original weight, was carried around, 72.158% was used or lost, 70.10% was fuel used.

(70.10/27.84 = 2.518) You must put in 3.518# of fuel at the beginning, burn 2.518# to have one extra at the end (to increase your range).

It is properly arguable that doubling the flight distance record was therefore 3½ times as hard aerodynamically and structurally for the Voyager, (impossible for any other plane) since 3.518 is the number that accurately relates the takeoff gross weight, return weight and fuel used on the Voyager World Flight.

Airplanes fly on an L/D, lift/drag, ratio. Lift must equal weight, drag is directly proportional to weight. Hauling fuel weight on a light structure is the problem, the task to be done on long range flight. Range and all efficiency numbers tie back into what weight of fuel you must add and haul to get one pound of extra reserve or range at the end.

If you add one pound of fuel at the beginning, you end up with .2842516# of fuel at the end, 28.425%, which is the reciprocal of 3.518 with the extra decimal points carried through in the calculator.

If you fly a perfect flight by the curve with no lost fuel or provisions, from 9694.5# GW to 3184.18# GW, 6510.32# of fuel, the numbers are better, 3.044# of fuel to have one at the end, a 32.84% yield.

This number juggling shows you that flying 3.518 times heavier, your drag and your fuel usage is 3.518 times greater, flying at the constant actual L/D ratio, which makes the fuel lifting and haulage and structural problem 3.518 times as difficult for the Voyager. Recognize for any other plane it is flat impossible. Only the Voyager, with its superior creative configuration, its superior composite structure, and its outstanding aerodynamic and propulsive efficiency can accomplish the task of world flight.

*This does not account for the crew weight loss.

They had made several records and Dick was looking for a proper challenge! It is always the same kind of people that are involved when the world makes great strides. In every case, the people are highly intelligent, creative, dedicated, internally driven, marching to their own drummer. They accomplish their objectives whether or not money is readily available. Money is never their objective, the task at hand is, the new, large step ahead. They can be sociable, but they are loners who listen to themselves.

When the Voyager lifted off from Edwards Air Force Base at 8:01:44 on Sunday, December 14, 1986 for its 216 hour, 3 minute and 44 second, 26,358.6 Statute mile conquest of the earth, it was the culmination of a 5 year effort.

Naturally there were those who presumed Voyager was the effort of more of history's promoters, charlatans or fools, but of course they would soon learn that like the Wrights, Lindbergh, Henry Ford, this was the genuine article.

Yes, it started on Burt's napkin sketch over lunch and quickly went through a configuration optimizing metamorphosis. It had few drawings and was designed as it was built, but every decision was professional.

People with money are money oriented and are loathe to give it up. Many wealthy corporations and individuals failed to measure up to the people who backed the Spirit of St. Louis, and there never was a real source of funding. Enlightened manufacturers in the Aerospace industry came forward and provided the necessary materials, equipment, engines and propellers that ultimately created this milestone of flight. Thousands of VIP supporters and well wishers contributing $100 each on a grass roots basis paid the hangar rent and provided sustenance for Dick and Jeana and Bruce Evans. A small core group of volunteers sustained the effort over the early years. A growing commune, mostly volunteers, provided the larger support group necessary for final testing and mission readiness. Burt Rutan and his homebuilt airplane business provided funds and facilities, the seed money that covered the entire initial construction phase.

The effort was so audacious; doubling the record, flying for nine or more days, circling the entire earth, taking on world weather in the heaviest, relatively lowest powered plane ever made, and flying it from a horizontal phone booth sized cabin that even much of the knowledgeable, well wishing aviation community thought they didn't have a chance.

Those who were unusually perceptive recognized that it was a brilliant concept professionally executed, professionally tested through 67 prior test flights with ever better engines, equipment, propellers and planning. Just like a spacecraft or a research aircraft, you go when you have worked your way through an engineering test program, found and fixed the problems, made it mission ready and that had been done just as professionally as in a proper space program. With creative people there was little structure, but formality is not a requirement, just correct decisions.

Not since the Wrights had the people doing it designed and built their own aircraft. Voyager was an outgrowth of the homebuilt aircraft movement. "Surely these must be amateurs and fools," some thought!

What the perceptive knew is that homebuilt aircraft have become part of the leading edge of flight. While the aircraft industry is dabbling in composites, Burt Rutan and the sailplane builders showed the world how to build magnificently efficient aircraft that are entirely composites -- with near perfect smooth contours that maintain laminar flow, that make a B1 or an F16 look like a crude effort on close comparison.

The Voyager is 340% more efficient than an industry produced two place private aircraft based on a hard numbers comparison! It will carry three times the load while going faster on a lower fuel burn. In the strict discipline of the laws of Physics, you only make a 340% improvement when you are very, very creative, smart and professional and have a very sophisticated grasp of the Physics book.

To carry the huge fuel load on a structure that at 938 pounds was less than 10% of the gross takeoff weight, requires a brilliant configuration concept that spreads the load over the entire aircraft to reduce the bending loads -- and then an optimum use of the best of modern carbon graphite epoxy composites. The plane had to be so heavily loaded, designed without excess safety margins, that if Burt made any significant design mistakes, or Dick, Jeana or Bruce made any significant fabrication mistakes, the crew was going to get killed.

These were not fools but very knowledgeable people accurately assessing the problems and the risks. Everyone who understood knew there were risks, but if you watched the wing deflection in test vs the design ultimates, if you knew how much side load the landing gear had taken during fueling deflection, if you looked carefully and found no evidence of the precursors of structural trouble, you had solid engineering evidence that the flight would be safe and would be accomplished, if we guided it out of heavy weather when it was still heavily loaded.

The people who designed, built and tested the plane had done a professional and perceptive job. A few small examples are instructive: If the exhaust stacks failed during the flight, there were cables to keep them from flying back through the pusher prop. Bruce Evans, Glenn Maben, Ferg Fay were pros.

An engineering oriented reliability investigation of the critical autopilot identified accurately that the artificial horizon, not the autopilot electronics or servos was the item most likely to fail. The horizon did fail, we had the spare aboard, Jeana was prepared to install it inflight and did. In a space program or a flight of a record aircraft, that level of readiness tells you when you are ready to go.

When Voyager lifted off on the World Flight it was a refined aircraft! It was on its second set of engines, its fourth set of propellers. There had been innumberable major and minor improvements, all aimed at improving its performance and reliability. Good work had been done, the plane was ready.

Voyager Mission Control under Larry Caskey, a flying pro who understood people and knew how to recruit a professional team had aircraft industry, weather and communication people that were pros in their own lines who were up to the task of guiding the flight from the ground just as well as the teams that made Mercury, Gemini, and Apollo missions successful.

The whole story will come out in Dick and Jeana's book and in Burt's book. Our task here is to tell you and teach you how to properly understand the performance aspects of the Voyager Flight and what really happened.

As you already realized, we are giving you little sub chapters to give you the right insights, to give you the right perception of what really happened. Now we will give you the narrative and the facts and the comparisons so that you can have a real grasp of just what happened on this milestone of flight.

Look over the log and the summary of facts on the bottom that will tell you where it went and how far it was and how long it took and at what speeds and at what fuel consumptions. In giving you the narrative and the facts and the comparisons in this book, it is our objective to give you the meaning and the insight and the understanding of what all the raw numbers mean and how you can correctly understand what really happened.

Voyager Mission Control was much more than a support group, it was an integral part of the flight. The central job of Mission Control was to keep Voyager out of heavy weather, guide it into the best weather and the best winds.

To do this task, Larry Caskey, the Mission's Operations Director, assembled a group of industry professionals with every bit of the competence and dedication of a NASA Mission Control.

Len Snellman, the Weather Director, retired from NOAA Headquarters Staff and one of the top weather professionals in the country, assembled a hand picked group that included some of the best weather professionals in the United States.

With DWIPS, developed by Clarence Boice, the weather team had full time access to the satellite weather images around the entire earth. The computer based DWIPS System allowed all of the recent world weather to be stored and brought up for a display at will, either individually or in sequence so that the weather movement could be watched. The accurate latitude and longitude grid on the television monitor allowed the weather location to be accurately compared with the airplane location. The system allowed a full earth view or the ability to zoom in on a localized area. Furthermore, it was capable of a visual satellite picture or an infrared image and with the use of the infrared image and a "mouse" individual areas could be probed for their temperature and therefore their approximate altitude. On infrared, the very high clouds showed up as pure white, the lower clouds varying shades of gray, all apportioned to their temperature and height. *

In addition, through land line hookups with the National Weather Service, selected world-wide weather, data and charts could be brought directly into Mission Control in normal printed and chart form.

Don Rietzke recruited an equally competent group of Communications Specialists, all of whom both flew and were Ham radio operators. Each had years of experience in making communications work and in hearing the often very difficult transmissions.

The Communications simply had to work or Weather and Mission Control could not do its job. Both HF, High Frequency long range radio was used together with land line links around the world, along with a separate UHF satellite uplink system, so that both satellite communications and HF/land line systems were available throughout the mission.

Communications often were very faint, especially from the other side of the world, but with one or two minor exceptions, always worked, a credit to the competence of the Communications Group.

*There were important gaps in the satellite coverage, most notably the mid Pacific and Central Indian Ocean. IR data was available from the Japanese satellite every three hours. The science and the computers never solve everything. It takes people with knowledge and judgment. The weather people earned their respect!

When after months of preparatory work, a family illness prevented Don Rietzke's participation in the actual mission, Dick Blosser, Walt Massengale and the entire Communications Crew took over and worked flawlessly throughout the difficult mission.

After a half year's preparatory work, the Performance Group under Jack Norris, the Technical Director, had investigated every facet of routing, weather, winds, as they correlated to airplane performance. We knew before lift off with absolute certainty that the numbers were there, that Voyager had the performance to circumnavigate the earth. We knew the potential variations in flight distance and winds and time and altitude, and their potential effect on the fuel reserve and the time of arrival.

We had the flight planned and precalculated before takeoff with backup computer runs of the potential variations, because we knew in advance that the inflight calculations could be treacherously complex and knew that if there were errors in the inflight reports, gaps in the data, or communication lapses, we would be in deep trouble, unless we already knew the possible range of answers. The plan was to have a team of volunteers actually doing the inflight calculations and to be in a position to judge whether the calculations could be trusted in comparison to the preflight calculations. The final inflight accurate checks would be the very accurate calculations possible as the tanks ran dry and the last tip tank fuel was used. When the inflight data produced errors, we saw it and were prepared to cope.

The Mission Control experience is one that will be remembered for life by everyone who participated. It was a singularly excellent group of people and absolutely everyone was working flat out to make the mission a success. Especially during the first few days, it could not have been more intense if it were a bunker running World War III. Absolutely everyone was intent on making the mission a success. Note clearly that the term was not tense, but intense. It was a professional group used to taking the heat and handling it. There was only one case where tension was a problem, rather amazing under the circumstances, and that was handled quietly and smoothly.

The rather amazing point that would otherwise be missed by history is that the final decision to go was made so quickly, much of the team coming from various parts of the United States was not even there at takeoff time. As a result, there was no time or opportunity to do the necessary last minute retraining and tune-up. The professionals simply arrived, slipped into their job and the problems and gaps that were occurring were corrected in real time with Larry Caskey's Operations Group coordinating all.

Everyone who worked at Mission Control was necessary and everyone performed like a pro. Naturally there were bonds of camaraderie and good feeling formed that will be lifeling for all of the fortunate individuals who got the chance to participate in the World Flight.

It was neither easy nor neat. It was nine grueling days with constant challenges and problems and for many, little sleep. The bottom line success of all functions justifiably qualifies everyone there as a pro.

It's proper here to give a special accolade to King Radio, who provided the excellent electronics for the World Flight of the Voyager. They provided more than the electronics and communication equipment. The full time autopilot was very important, the VLF navigation system and new global positioning satellite system allowed the Voyager to always pinpoint its position, therefore allowed us in Mission Control to accurately compare the position of the aircraft and the weather that we could see on the DWIPS television grid. Voyager was the first record craft in history to be guided by pinpoint navigation, coordinated against real time world-wide weather by a totally professional team. It should also be noted that Motorola provided the special military communication packages that permitted UHF satellite uplink for a backup of round the world satellite communications.

The special King autopilot developed by Jim Lehfeldt of King is worthy of special note. The great flexibility and heavy load of the Voyager coupled with the configuration that made it successful, caused it to be dynamically unstable above 82 knots. It could not be stabilized by aerodynamic or simple bob weight systems, but required a high technology autopilot with a pitch rate sensor that gave a leading control input, a correction that led by $\frac{1}{4}$ cycle or more before the actual oscillation in the pitch axis. The excellent autopilot worked perfectly, but was destined to have one inflight mechanical problem.

Looking at all options on stabilizing the plane, Jack Norris, an expert on spacecraft controls, concluded that the autopilot simply had to work. The rugged control servos used on the heavy, faster Cessna Citation were not likely to fail, and the reliable electronic package could be expected to succeed. Discussions with Jim Lehfeldt concluded the basic artificial horizon was actually the component most vulnerable to failure. Norris and Lehfeldt agreed a spare would be carried. The horizon did fail, Jeana was prepared and successfully changed out the horizon in flight. In the complex aerospace field, there is the potential for a lot of problems. The score is kept not on whether you have problems, but rather on how well you anticipate and handle them. Lehfeldt, a pro, knew his system and we nailed the right item.

Mission Control under Larry Caskey did a great job. The communications were often extremely difficult. Dick Blosser, Walt Massengale and the gang came through like pros. Len Snellman's Weather Group with individual stars like Rich Wagoner, Larry Burch, everyone in the group were literally life savers and were fundamentally necessary to the safe return of the Voyager.

Everyone involved is worthy of note and should have their story told. Space makes that not possible of course, but every name is shown on the attached list. As a proper representation, we'll single out just one who came late. The memorable way in which he answered the telephone best exemplified the special flavor of Voyager Mission Control: "This is Milton Mersky, how may I help you?"

The word best describing Mission Control people was "Genuine". They were all there because they selflessly wanted to help, each a pro in his own way, and everyone came through when the heat was on.

VOYAGER MISSION CONTROL

LARRY CASKEY	DIRECTOR	MISSION CONTROL
Len Snellman	Weather	Director
Don Rietzke	Communications	Director
Jack Norris	Technical	Director

OPERATIONS

Director Larry Caskey

Bob Brubaker
Martin Caskey
Gilbert (Gil) Fortune
Isabel Fortune
Fitz Fulton
Mike Hance
Clarence (Cobb) Harms
Jim Kunkle
Mike Melvill
Milton Mersky
Conway Roberts
Don Taylor
Jim Whitman

WEATHER

Director Leonard (Len) Snellman

Richard Barrett
Larry Burch
Fred Meir
Steve Mendenhall
Ken Modlin
Pete Mueller
Walt Rogers
Frank Smigielski
Lynn Snellman
Larry Tonish
Hector Vasquez
Rich Wagoner
Mary Wagoner

MEDICAL

Flight Surgeon Dr. George Jutila
Flight Nurse Suzie Bowman

COMMUNICATIONS

Acting Director Dick Blosser

Dave Bearden
Mike Downs
Stanley Dusza
Rich Elder
Stuart (Stu) Hagedorn
Greg Kordes
Walt Massengale
Murray Olson
Don Rietzke
Robert Sechrist
Dick Shane
Allan Siebert
Edward Sindeff
Larry Stencel
John Swancara
Chuck Whittington

PERFORMANCE

Director   Jack Norris
Asst. Director Brian Hobbs

John Burns
Jim Johnson
Eric Knutson
Milly Norris
Jim Richmond
Kelly Sandfer
Terri Smith

Brent Silver   Computerized
Aerodynamic and
Flight Analysis

Doug Shane, Chuck Richey, Mike Melvill and Fitz Fulton of Burt Rutan's Scaled Composites were an integral part of Mission Control and a continuing help.

Keith Gordon of QANTAS provided crucial insight on ITC, monsoon movement and tropical winds, the hard data that permitted optimum route planning.

THE HANGAR CREW

THE PEOPLE WHO CAME BEFORE MISSION CONTROL AND WERE THERE FOR THE LAUNCH

Bruce Evans - Crew Chief - #1 Man who built the Voyager with Dick and Jeana from the start. Deserved and shared the awards!

Mike & Sally Melvill - Earliest Volunteers - Dick's flying companion, First builder of a Rutan Plane.

Chuck & Joan Richey - Early Volunteers - Landing gear design - Continual Supporters like Sally Melvill

Lee Herron & Diane Dempsey - Early Volunteers - Weather chase pilot - Building Helpers, PR - Ground chase and control

Ferg Fay - Front Engine - Retired from Air Force and Rockwell Space Program - Been there and done it. Knew the long view.

Glenn Maben - Rear Engine - Aeronautical Engineer, Aerobatic Pilot, Hands on know how

Neal Brown - Do everything man - The jewel who helped everyone on everything.

Gary Fox - The swing man backing up the crew.

Kevin Furman - The fine young helper who did anything necessary.

THE OFFICE CREW

Peter Riva - Media Relations - Business Management - A driving force over the goal line.

Wanda Wolf - The sharp, attractive telephone operator, the voice of Voyager.

Kelly Chandler - The all purpose Kelly girl - Doing whatever was required.

Terri Smith - Jeana's best friend who helped at everything, even the final flight plan.

Suzie Bowman - Flight Nurse - All around helper and adviser.

Dan Card - The pre flight General Manager of inside business.

Gary Gunnell - The post flight General Manager of inside business.

Irene & George Rutan - Mom & Pop - They started it all, helped bring the public aboard.

Lee & Frances Yeager - Jeana's Dad & Stepmother - Helped the growing public.

Evaree & Doug Winters - Jeana's Mother - Supported the Volunteers.

Dale & Billie Fox - Parents of Gary Fox - Nice folks, all around helpers.

Sylvia Jutila - The Flight Surgeon's wife and ace chef.

There were many others who helped along the way. It was a growing living commune supporting a dream that was to become history.

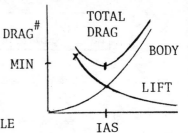

THE INNER SECRETS OF LONG RANGE FLIGHT

READ THIS - THOUGH MORE TECHNICAL, IT'S UNDERSTANDABLE

AND IT WILL TEACH YOU THE REAL SECRETS OF LONG RANGE FLIGHT

Drag (pounds) x Distance (ft) = Energy (ft #) $\cong$ Fuel Qty(Gallons)(BTU or ft #)

If you want to fly around the world, the distance is fixed, so to minimize the fuel load you must lift and haul -- you must minimize drag, since drag times distance (ft #) is energy, equivalent to gallons of fuel.

Naturally to minimize drag, you must have a slick low drag plane, but there are two less obvious points hiding here that are the real keys to long range flight. First, you must fly SLOW, since form drag increases as the square of speed. Twice as fast, four times the drag.

Second, you must understand that there is drag due to lift, an item of huge importance if you must lift and carry a huge fuel load. To fly slow, you need a big, large area, wing, but of even more significance, you need a very long high aspect ratio wing with a small chord, because theoretically a wing of an infinite length has zero induced drag, the drag force due to lift.(You can see that if you make a very long wing, you inherit a huge structural problem. You must create a structural wonder to produce an aerodynamic wonder.)

Few people realize that drag due to lift decreases as the speed squared, twice the speed, 1/4 the lift drag! But this means to slow down, lift drag increases, so you get a huge drag due to lift, unless you have a huge aspect ratio to minimize lift drag flying slow! The Voyager aspect ratio, span/average chord, is 33.8.

Minimum total drag is found at the speed where the two drag curves cross and are equal.   You get the slowest flight and thus the lowest body drag and equally low lift drag by having a very long high aspect ratio wing that gives low lift drag at a slow speed.

You thus need to create the longest and relatively most highly loaded and lightest wing ever created -- and then you must deal with the extreme flexibility that it will surely have. Slow is as difficult as fast!

The Hercules Magnamite Carbon Graphite fiber/epoxy spar and the Carbon fiber/epoxy Hexcel sandwich skins created arguably the greatest wing ever made, a true structural wonder. Everything else was built on that foundation!!!!

You have just created the basic configuration for a very high lift/drag ratio, the first and most fundamental key to a long range flight. If you use the optimum airfoils and body shapes, have slick flaw free surfaces that maintain laminar flow, you have created a maximum L/D aircraft. You must fly it at the magic speed where the two drag curves cross, where the minimum drag is found. Of course you must also have very efficient engines and propellers.

$$\boxed{\text{Breguet Range Equation}} \quad \text{Range} = \frac{\text{Prop Eff \%}}{\text{Engine SFC}} \times \frac{L}{D} \times \text{Ln}\left|\frac{\text{T.O. Wt\#}}{\text{Land Wt\#}}\right| \times 375 = \text{Statute Miles}$$

$$\text{Range} = \frac{.8}{.4} \times (28.3+)^* \times \text{Ln}\left|\frac{9694.5}{2683}\right| \times 375 = 27266.3 \text{ Statute Mi}$$

(Min. Curve)*

*(Voyager L/D is $36\frac{1}{2}$ with laminar flow, 28 without.

You would like to have an L/D close to 32 so you have some margin for off optimum flight, loss of some laminar flow due to bug contamination, the practical type considerations, the realities of a challenging world flight.

The Breguet Range Equation (pronounced Bra Gay, as in French) shows you that you must have a huge fuel load carried on a very light structure, so that the takeoff weight ratio over the landing weight is almost 4.

With engines, landing gear, propellers and equipment and crew and provisions, the structural weight must be only 10% of takeoff weight, a task that was simply impossible with metals before modern composite structures. Voyager is first of all a structural wonder.

Breguet shows us that if we finally have efficient propellers and engines capable of operating over a broad power range, yielding a specific fuel consumption close to .4 pounds of fuel per horsepower hour average, together with our high L/D, big fuel weight and light dry weight, we have all the basic elements for world flight.

As you will see, we must finally fly the plane fast at first when it is heavy with high power and fly it slower and slower as it grows lighter (at a speed proportional to the square root of the gross weight), always at the same angle of attack to maintain the optimum minimum drag speed at all weights. The Continental engine and Hartzell propeller efficiencies were absolutely essential to the flight with its broad speed/power range.

Long range airplanes like Voyager fly at the same optimum L/D ratio, always at the same angle of attack. At takeoff when they are heavy, their lift must be high so they must fly faster, and the drag is proportionally higher so the power requirement is quite high, since both the drag and the speed are higher. At the end they can fly very slow and just sip fuel. Voyager uses up fuel fast at first, heavy and travels three times as far on a pound of fuel at the end when it is light.

Voyager is first a structural wonder -- then an aerodynamic wonder and you will see later that it is flown to a speed vs the square root of the gross weight profile that gives it a specific range of 2 N.Mi/Lb of fuel at first and a potential of 6 N.Mi/Lb at the end.

YOU NEED	Hi L/D	Hi T.O. Wt Lo Landing Wt	Prop Efficiency Low Engine SFC	+ Fly Max L/D Speeds
1	Low Drag High L/D Airfoil	Fly Slow Long Wing Laminar Airfoil	Low Body Drag Low Lift Drag Smooth & Right	
2	Super Strength Super Light	High Fuel Load Low A/C weight		
3	High Prop Eff Low Engine SFC	Broad Range # Fuel/Hp Hr		
4	Fly Max L/D Speed	$V = K\sqrt{GW}$		

28

This analytical comparison has many numbers. You can screen out the numerical minutia as you read and get a good feel for the flight.

THE REFINED LOG OF THE VOYAGER FLIGHT
A COMPARISON WITH THE FLIGHT PLAN

The Hasty New Plan

The flight plan for the new Northern route, using the Lockheed Data Plan computer terminal, was hastily put together in just three days when the decision to launch was made. Terri Smith, the least experienced person in the group, jumped in to help in that hectic period and her plan not only worked, it worked spectacularly well. When the flight arrived at the 75° West, Caribbean decision point for the final route home, the flight was just 16 to 18 pounds, 3 gallons, off the plan. It was the detailed plans generated over four months for the Southern route, now suddenly obsolete, that made this possible.* The final plan, which would take two weeks to complete, had a few obvious errors but proved to be a first class skeleton for the flight, if we were all just smart enough to understand it and believe it.

With no experience in the field, Terri, Jeana's girl friend and riding companion, came through with flying colors, under pressure, when the others, under pressure, were working out how to cope with the new decision to fly a "decide as you go" C_L, thus undefined speed changes, a task that took 15 computer runs, a stack of data over a foot high, to find the limits of the possible effects at all altitudes up to 20,000 feet. The complex planning challenge was now even worse.

Like all Voyager planning, Terri's used 6700 pounds of fuel, a 2600 pound plane, for a 9300 pound gross weight. The plan gave a 571 pound reserve after a 22,046 Nautical miles, 25,370 Statute miles point to point "geometric route". The flight took 223 hours and 44 minutes (9 days, 7 3/4 hours), at 5000 ft. altitude, 6500 ft. density altitude at .5 C_L speeds. (The standard Southern flight had been 22,702 Nautical miles, 26,126 Statute miles - 755.9 Statute miles further - took 232 hours and 27 minutes, 9 days, 16 hours, 27 minutes and returned with 500 pounds of fuel.) (See rear inside cover for the actual flight plan.)

Extra Miles, Extra Fuel

There would be extra miles on any flight compared to the simple geometric route. We knew there would be 600 to 1000 miles extra to avoid weather and undoubtedly some climbs to avoid weather, especially on the Northern route. Though the Northern route was shorter, it was not expected to "play shorter" against weather. The final number was 859 extra Nautical miles, 988.5 St. Mi.

We expected that the ground crew would add extra fuel at takeoff, but how much would be a last minute decision, and in the crush of the launch, it was two and a half days into the flight before Performance was able to locate the data, lost in transit, that showed 311.5 pounds extra fuel had been added and that the dry plane was 83 pounds over 2600 with final provisions for a 9694.5 pound T.O.G.W. (9300 + 311.5 + 83).

*Four months hard work by Jim Richmond with Milly Norris running the Lockheed Data Plan, made routing, winds and performance an open book. The standard summer flight plan tailwind, 8½ knots, was within .007 knots of the actual flight!

<u>Getting to Front Engine Shut Down</u>   (+642 N Mi)   (5:55 Longer)   (311.5# Ahead)

The addition of the 311.5 pounds of fuel provided an initial extra pad of just 642 miles at 2.061 Nautical miles per pound, until it was down to the 4400 pound fuel remaining point where on the planning flight the gross weight was at 7000 pounds and the front engine was shut down.

Amazingly,* on the actual flight, (after one too early attempt at engine shut-down at a too heavy 7598 pounds) the front engine was shut down 83# past 4400# accounting for the extra provisions, at 7000.7 pounds, just .7 pounds off. The shutdown succeeded for 1:20 until weather forced a restart.

Of greater importance, the 311.5 pounds of extra fuel gave a pad for extra mileage, climbs, etc. that made <u>direct</u> comparison possible between the flight that had more early and late miles, climbs, varied winds, varied altitude, etc. vs the plan.

Before we now start to make the specific comparison, one word of caution is appropriate. The best insight is gained by making the broad overview comparison -- <u>not</u> getting too deep into the minutia where things get too deep, too complicated and the complex calculus like problem done be many, many individual calculations with constantly changing weight, speed, fuel burn, with speeds that also change with winds, and then also as altitude varies. The precise calculations are a snake pit of numbers for the unwary or the experienced that vastly exceeds the interest and attention span of just about anyone. We will avoid that mistake here. A specific engineering analysis follows.

As above, the 311 pounds of fuel carried the initially heavy plane only 642 miles farther at 2.06 Nautical miles per pound to the equivalent point where 4400 pounds of fuel was left. Note that equal fuel remaining is the comparison, and that <u>fuel remaining</u> will be the comparison carried forward in this analysis. The plane flew on until 83 pounds more fuel was burned off equivalent to the 83 pounds of extra cabin weight and the front engine shut down amazingly just .7 of a pound off the 7000 pound target. Shut down with the extra fuel came at 52:41 vs the 46:46 plan with less fuel. After 1 hour 20 minutes however, weather forced an engine restart at 54:02 and it was <u>shut down finally at 58:41 at Bohol Island in the Philippines</u>, North of Mindanao, West of Leyte. Significantly, the engine had run essentially an extra half day, just 5 minutes short of 12 hours, except for off times of 5 minutes and 1:20, thus exactly 10:30 extra.

<u>Bohol, Philippines</u>   +307 N Mi   1:10 Behind   +20#

At Bohol, Voyager had traveled just 307 Nautical miles farther than plan due to detours and significantly with the extra half day engine running was <u>not</u> 642 Nautical miles ahead, but just 307. Again amazingly, the flight <u>log</u> shows 4078.5 pounds of fuel remaining where the plan shows 4078, but the checkpoint had moved farther, 1° of Longitude, 60 Nautical miles, 20 pounds of fuel at 44 #/Hr getting to shutdown so 4058 pounds by the plan is in

* It was amazing because we were developing a serious gross weight error, and operating with wrong data Dick and the plane did exactly what Burt's Curve predicted. 7000# had always been the nominal shutdown weight!

fact the equivalent comparison. The plane thus was really 307 miles and 20 pounds of fuel ahead at Bohol. It arrived at the Bohol checkpoint 1° farther, at 124°East at 58:41 vs the plan of 57:04 for 125°02'E about 1:10 behind the clock, corrected 27 minutes for the Longitude difference, also corrected 60 Nautical miles and 20 pounds of fuel, but having traveled 307 miles further with favorable winds.

Hatyai, Thailand   +352 N Mi   3:00 Behind   -62#

Right after the transition to single engine heavy is the toughest time to meet the performance curves and the fuel performance started to fall behind. Because of the developing problem of knowing the correct gross weight, the plane was beginning to be significantly heavier than the crew thought, 544 pounds at shutdown. The plane was going significantly too slow to get its best performance. Thus, by the next checkpoint, Hatyai, on the South Thailand Peninsula, the fuel remaining was down to 3638 pounds vs 3700 on the plan, 62 pounds behind at 72:44 vs 69:44 hrs, 8668 N Mi vs 8316.

Sri Lanka   +371 N Mi   2:36 Behind   -49#

Slow and below curve fuel performance continued but the winds were a bit favorable despite some unfavorable weather areas and the crew arrived at Sri Lanka a bit better, 49 pounds behind the plan. Again the plan fuel had been adjusted 168 miles, 45 pounds of fuel for a valid comparison with a checkpoint 2°48', 168 N Mi sooner at 83:11 (+1:39) vs 82:14, 9763 N Mi vs 9560 (+168).

Mogadiscio, Somali   +448 N Mi   4:16 Behind   -133#

Now due to weather and the need to fly over the volcano that is Sri Lanka, a climb costing 42 pounds of fuel was involved and the gross weight discrepancy was growing, thus slow flight and below curve fuel performance is clearly seen on the accompanying specific range curve and Voyager arrives at Mogadiscio, Somali, the next checkpoint, 133 pounds behind plan with 2667 pounds vs 2800, its greatest deviation from plan but less than 2% off, since 70 pounds is 1% of 7011.4 total fuel. Weather had not been friendly over the Arabian Sea and the winds had fizzled out adding to the problem. The time was 107:43 vs the plan of 103:15, 4 hours 28 minutes behind, but the plane had been flown 448 Nautical miles farther, 12,110.6 vs the plan 11,644 (11,625 adjusted for Longitude) just about what would be expected with the extra miles. The too slow indicated airspeed was not hurting time because the higher altitude being flown made the true airspeeds higher nullifying the slow flight effect.

The numbers shown are corrected, 5 pounds of fuel and 12 minutes for the 19 Nautical mile difference in checkpoint location.

Dick had done some circling, but that would be properly accounted for in air miles calculations. Ground miles would also be correct. The winds were slightly better than those calculated to make up for the circling done.

31

THE FUEL TRANSFER SYSTEM GROSS WEIGHT PROBLEM

That was how it was at Mogadiscio, but it was not quite so neat in Mission Control that day and night. Fuel consumption from the engine flow transducer looked right but gross weight and fuel remaining, looked far too low, dead wrong!

From months of hard work and planning, Performance knew with certainty that if things were right there was without question the fuel to make the flight and with targeted margin. Reports from the plane however, were terrible. The gross weight and the fuel was falling rapidly according to the on board fuel log* and disturbingly the plane was said to be just barely flying. Based on the planning work, we had to be okay, but the gross weight log shouted "Trouble". Could there really be a problem? We had to face the possibility. A simple look at the flight plan showed the plane GW reports to be 1000 lbs low and the problem was now clearly getting worse faster and faster.

Could fuel be leaking out? Could the engine flow look right but be off and the engine also going bad, an unlikely double failure? If the plane was flying like a sick bird, however, if you were objective, you had to at least consider the possibility of engine trouble. The problem was that Dick was the guy that needed the help, but only Dick was there to judge if the engine was sick or sound, the power fading or not, or could there be damage to the plane and lost fuel. The Coast Record Flight should have found a transfer system read out error if there was one. (On that flight, fuel, after transfer had been shut off tight!!!)

Finally, painfully, we asked Dick what we didn't want to ask. What did he think? Was the engine acting like it had a sick valve or some other malady? Dick, bone tired by then as we all were, clearly wanted no bad news, but we found he didn't think there was an engine problem, a question we needed to eliminate.

Our plan already said what the numbers should be, but could we believe that they applied. The newly recruited and fortunately very sharp Performance volunteers were busy calculating and back calculating the flight and the numbers agreed within the expected errors. Performance correctly believed Dick's reported fuel burn numbers would be a few percent optimistic.* No problem now, but a big potential problem at the end. The data was not accurate!

We all saw Dick, dead tired, needed encouragement. Weeks before the launch, we had seen that Mike Melvill was the trusted friend to be on the radio when the fat was in the fire and that time was now. Everyone blamed the transfer system, but there was no proof and only Dick knew how the plumbing of the fuel system had been done. The schematic was different than the plane.

Chuck Richey, Brent Silver, Burt, were all calculating and being tired, we encouraged this welcome check and extra brain power. A wise manager knows a smart group will always beat the individual because all the best varied ideas come out, but it was Performance's job to not let bad information out to Dick.

Mike was put on the air to send good news. We got Mike aside however, specifically concurred that he should send good news to fix the human equation but then told him to not hold back at all but not to go overboard, that the group number most probably was a couple of hundred pounds high. We told him that was not a large problem but that the real test would come later. Would we get the long legs at the end of the flight? What decisions and problems would be encountered subsequently? We needed to give Dick reason to relax and have

*There are two systems, "the out of tank" Fuel Log used for Gross Weight calculation, and fuel flow to the engines which can give a double check. The Fuel Log proved to be 1000 pounds wrong! The Fuel Burn was biased by weather, a trap!

confidence but could not afford to send him from 1000 pounds too little
fuel to too much. Fortunately on that very tired night, the advice was
absolutely correct. Richey's welcome excellent log paper check was initially
200 pounds high, but was soon to be corrected and then be right on.

Dick noticed bubbles in the fuel line <u>flowing the wrong way</u> and concluded
fuel had been flowing back into the tanks. At the same time on the ground,
checking with old friends, we established the six ways the turbine flow
meters could read incorrectly. Air in the lines was the solution put forward
by the experienced instrumentation engineers. Significantly, however, they
pointed out that turbine meters incorrectly <u>counted positive counts</u> even
<u>when the flow was backwards</u>. Dick, working with Jeana's back calculating
work, convinced himself before Kenya that fuel was indeed normal. The flight
log says good news, but to both Dick and the ground people, relief from the
strain of that problem was much more than just good news.

Simply flying the plane now at a higher speed for the corrected gross weight
promptly cured the sick flight characteristics. We had simply been trying
to fly the plane far too slow at the incorrect indicated lighter weight.

Now here comes the more subtle point. <u>If you carefully calculate</u> the fuel
using all the reported points, you do <u>get a wrong answer</u> on fuel used, as
we in Performance suspected. We didn't trust the data was accurate, believing that it being reported when things were going well would be a few
percent optimistic. There were clearly a lot of bad times. That again
turned out to be right on target. The raw data log calculation shows the
fuel numbers were 2.69% optimistic on the amount actually used. An unimportant error in the middle of the flight, but <u>a large, unacceptable error at</u>
<u>the end</u>, when the fuel left is small, where great accuracy is needed, since
it is small, but it will take you three times as far. The raw data log
comes out <u>178 pounds high at flight end</u>, a big error compared to the 106
pounds left. Performance knew what <u>not</u> to believe - <u>but we couldn't be sure!</u>

Now here's a very interesting and fortuitous trick of fate. Dick and Jeana's
numbers calculated tired under cramped terrible conditions were 118.8 pounds
low compared to the raw data log. He said 5257.6 pounds gross weight vs
5376.4 on the computer calculated raw log. However, the refined and corrected
log says 5259.9 pounds. Dick and Jeana were within 2.3 pounds of the best
and final computer predicted number. Performance's check number coming forward as carefully as possible was 50 pounds below the erroneous raw data log,
74 pounds above the refined log 5031 pounds vs 4957 at 18°E. (A 1% error.)
On a sleepy night we were groping for the mid point and missed by just 12#.
Some people wondered if Performance really knew what we were doing because
we were not rushing forward with a new number every time someone else calculated a new number <u>so we published the 5031 pound report at 18°E that</u>
<u>had been sent up to Voyager, signed it, dated it and told everyone they</u>
<u>could bet their socks on it within the accuracy that the numbers would produce</u>.

We knew what everyone else didn't, that if the plane really was working
right, that unless something bad happened, the plan would prove correct.
No one else realized the even more complicated problem we were trying to
keep track of. The altitudes were higher and that greatly affects speed,
time and fuel burn <u>rate</u>, and of course the winds were different. Performance
succeeded, because before the plane was home we told everyone, that average
winds had been normal, <u>the early return was primarily due to flying higher</u>.

TO: VOYAGER I

FROM: JACK

Don't be mystified any more -- Use 5,031# GW, 2,348# fuel at
E18 Longitude. 5% above "How Goes It". Expect transfer
transducer reading high. Can be air accumulation or five
other causes. Under control, with insight. Keep good
performance report coming a few times a day and you will
get home with good support.

Jack Norris

The African Adventure

Doug Shane, in a Beech Baron met the Voyager over Sumburu South, Buffalo Springs, a picturesque asphalt strip complete with elephants, Northeast of Nairobi. A climb test of the Voyager agreed with Dick's gross weight conclusion. The Voyager fuel was okay and the wing tips were okay too.

Once the gross weight was corrected over Somali, the plane still flew slow above its apparent .55 C_L optimum, but no longer below the .64 C_L climb speed, the big difference that had hurt performance since front engine shut down and the performance started getting better. In addition, the first really significant wind effect change began to come into play.

Studies done by the French just after World War II, which QANTAS provided to us, showed a semi permanent Northeast, Southwest flow convergence line diagonally across the waist of Africa and it is typical to find headwinds in the last half of Africa and the early Eastern Atlantic from 30°E to 5°W, as much as 2100 Nautical miles.

While African weather was just terrible, the convergence winds were <u>not</u> there for Voyager's passage, so Voyager would begin to get ahead of the clock -- but would suffer 1900 miles of equivalent winds off Mexico after the Gulf of Tehuanepec. With our planning studies, when the African winds weren't there we knew it and congratulated Snellman on his power in providing a special African wind gift that would allow the flight to deal with the anticipated Mexican winter winds, not as an addition to the standard plan, but as a swap for the African variety.

African weather was terrible, solid walls of cu, struggling over notches in frightening buildups to have them close in behind you, a flight over downtown Entebbe, not where we wanted to be, continuing bad daytime weather and bad nighttime weather over the remains of the African rain forest, no place to be in trouble. But despite all this and the fuel costs of climbs to 20,500 feet, we were back up to meeting the specific range curve, the winds were with Voyager, not against it and Performance was now gaining on the plan, even with Kenya and Zaire climbs that brought the climb cost to 100 to 110 pounds of fuel. Dick was exhausted, Jeana flew the dark of night !

Cameroon	+451 Mi	1 Hr Ahead (5:16 Africa Gain)	-50#

When the plane arrived at 12°E at Cameroon forced by weather 3°:48'North of the equator rather than 4°:9'S, no surprise with the ITC trying to deprive us of the "belted waist" of Africa, the Voyager's fuel was 2193.8 pounds, just 50 pounds behind the 2243 of the plan -- The other side of the world, terrible challenges and within $8\frac{1}{2}$ gallons of plan! The Voyager was now at 126:44 on the clock, now <u>exactly</u> 1 hour ahead of the 127:44 plan at the 14,163 Nautical mile mark, 451 miles more than the 13,712 mile plan.

Back at 18°E working with only air to ground fuel burn data and no climb data the Performance Group's official published inflight tally of GW was 5031#, 50# <u>below</u> the final complete raw data log of 5081 and 74# <u>above</u>(1%)the final refined log of 4957 at 18°E. We expected the reported data to be 2-3% optimistic and it was,2.69%. We approved a number in between having incomplete data. We could have been even closer than 1%, by just using the plan, 50# off.

Now so that this appraisal is kept honest let us reiterate what we explained before. Notice that we're saying that the flight is per plan when extra fuel had been added. As explained, this is not a conflict. The added fuel simply made it possible to do a direct running comparison despite the fact that extra miles as expected were encountered in the early flight. As expected the extra fuel was quickly used up in getting started heavy at only 2 Nautical miles per pound and with extra 2 engine time heavy we quickly lost half of the advantage gained, quickly back to just 307 miles ahead of the point to point geometric route plan. Without extra fuel, we would normally simply be cutting into the 171 pound margin between the calculated 571 geometric route reserve and the true 400 pound targeted reserve and would expect to be falling initially behind the nominal plan and then gaining some of it back as the plane grew lighter than plan. Recognize that 171 pound margin at the end of the flight at 6 Nautical miles per pound is not less but more than the initial 311.5 pounds of fuel at the beginning which only yields 350.8 N.Mi on an exact analysis.

With the initial extra pad of fuel and the fortuitous way in which the flight developed, it is comparatively easy to explain and understand if we pay attention, but in fact the simple explanation masks the very complex calculus problem that the flight of the Voyager is, where a pound of fuel is not a pound of fuel but rather an ever changing commodity where gross weight, speed and fuel burn are ever changing, further changed by winds and altitude effects. This simple explanation masks mind boggling complexity !

This flight was a real bear to keep track of in real time because it constantly fought the ITC, constantly had challenges and ended up following a highly variable speed profile. Fortunately, one of the most predictable things about the flight after you understand it,is that the plane was ① either going to make the minimum curve of the specific range curve and arrive back with predicted fuel, or as occurred,② go extra miles, climb and fight weather, underperform to a manageable degree, all within the capability of the relatively small help of the initial fuel added and the substantial capability of the transcontinental range reserve that the numbers predicted. That is the real way to understand what happened. If you get that, you know what Performance knew when everyone was buried in a morass of numbers. ✓

Out of Africa - To Two Emergencies

The log shows it was a teary morning out of Africa as well it should have been, after 2200 Nautical miles of frightening weather challenges. There was relief. A now easy trip home was envisioned but that was not to be. Though the log showed that oil was checked, somehow there had been trouble. The oil got low, it foamed, oil pressure was lost. Within minutes, everyone was on duty in Mission Control and Dave Mayrose, the Continental Project Engineer was out of bed and on duty in Mobile, Alabama. More oil was added. After agonizing minutes, the engine slowly worked its way back to normal, defoaming as it went. The engine didn't fail, another crisis had passed and that was fortunate indeed because Snellman's weather group had just spent the better part of three hours threading the needle, guiding the Voyager out through bad weather off the Western Coast of Africa. A retreat to land was not easily possible. Mobile's new synthetic oil could well have saved the mission right there.

Though few people understood it, the large Western mass of Africa pulled
the ITC Northward, <u>North</u> of the Voyager's path and the second Atlantic
crisis was the Voyager being flipped vertical as it inadvertently flew at night
through a towering cu at 33°07'. With limited roll control, Voyager fortunately came down right side up. An immediate course change <u>South</u> and grand
detour brought the Voyager again out of crisis. Despite all this, the winds
in the bad weather area had been very favorable and the Voyager arrived at
the close checkpoint, 30°W Longitude, 53 pounds <u>ahead</u> of the fuel plan, 1715.9
vs 1663 (corrected from 1666 for 15' Longitude difference).

It was now at 16,740.1 Nautical miles, 525 Nautical miles past the 16,230
plan with a $\frac{1}{4}$ Longitude degree, 15 mile correction and with a big wind and
at a significant altitude effect, an amazing 7:45 ahead of plan 148:41 vs
156:20 with a 6 minute correction for 15 minutes of Longitude.

<u>On to Trinidad</u>   +493.3   9:46 Ahead   +20#

Now finally the crew would get some break from the constant challenges.
There was still weather off South America but it was not African weather
and the Voyager arrived at Trinidad with 493 extra miles, 18,918.3 vs
18,425 and a whopping 9:46 hours ahead, 169:08 vs 178:54, but now only
20 pounds of fuel ahead, 1284.6 vs 1265 per plan. The miles decreased
a bit because we had come across the Mid Atlantic, North of the Equator,
until detouring just South to 1°45'S, rather than the 8°S that had been
planned to stay away from the weather to the North caused by the bulge of
Africa into the Atlantic. Only long after the flight was it remembered that
the plan had actually anticipated the weather to the North that caused the
frightening 90° bank. Four months work showed us the weather patterns.

Continental's sensational new liquid cooled rear engine had been running
full throttle since takeoff, with power modulated by lowering RPM on
Hartzell's great constant speed propellers. (That, amazingly had been delivered
as a completely new design in just two weeks by feeding John Roncz's computerized design through a conversion program then into the blade
milling machine. Blades are actually cut in less than 10 minutes on a CNC mill.)

Performance was anxiously awaiting the chance to see the really long legs
of the plane at the end when it is light, where the Nautical miles per
pound can actually go above 6, actually to 6 3/4 on engineering tests and
to less than 2 gallons an hour, less than half the fuel of a 2 place Cessna
or a classic Luscombe. We would get one quick look at long legs after
Trinidad, but then the decision to come home rich and safe would change the
fuel picture completely.

Before we reached Trinidad, the ever decreasing power requirements forced
the engine RPM below 2000 and after one leg at 1850 RPM off the mouth of
the Amazon, at 2°44'N Dick increased the RPM and the speed and for the first
time moved above the original .5C_L cruise curve to 80 knots for what would
be the beginning of a purposeful decision to fly home faster, and then,
conservatively richer, and finally on two engines. The fuel flow, which had
been comfortably within the performance curve dropped just below it at 80 knots.

Just before Caracas,Venezuela at 66°W, the throttle was reduced for the
first time in over 7 days to 19"Hg, 1820 RPM and the speed slowed to 68 knots
(.58 C_L) and the Nautical miles per pound jumped to 4.9 near the upper performance curve. Just after Caracas at 67°36', the throttle was cut to only
18" and 1900 RPM and the speed at 70 knots was at the .55 C_L that from data
appeared to be the optimum and the Nautical mile per pound went to 5.0 just
under the maximum performance curve, our best look at long legs on the
flight.

With the <u>high</u> lead of the "100 low lead" fuel after more than 7 days, the plugs
seemed to be fouling. At the next checkpoint, 74°5', the decision would be
made to come home less lean, at higher RPM and take it easy on the engine.
Lead fouling was a major concern. The Champion plugs had come through!

That would be the major decision of the flight because as shown in separate
calculations 190 pounds of the reserve which was heading accurately for the
targeted 400 pounds would be used up. After losing 109 pounds by a leak in
the left tip fuel cap, we would end up with 106.14 pounds and make a flight
that was dead nominal look close to the world. (190 + 109 + 106.14 = <u>405# vs 400)</u>

We arrived at 75°37' almost exactly at our 75°W Caribbean decision point
with 480 extra Nautical miles, 10 hours and 18 minutes ahead of schedule,
both because we had flown higher and had not yet encountered our headwinds
and with just 18 pounds of fuel more than the flight plan.

So here we were 19,783 Nautical miles out, 22,766 Statute miles, 86% of the
flight and <u>18 pounds, just 3 gallons off.</u> Fantastic !

351 of those 480 miles had been free, provided by the extra 311.5*pounds of
fuel added.   At first it seems incorrect that we had burned the correct
amount of fuel, though we had arrived over 10 hours early. That is much
more logical than it looks. The flight had been planned at an optimum
5000 feet, about 6500 feet density altitude whereas the average actual
altitude due to the weather was over 10,500 ft DA. The effect of this nominal 4000 foot altitude difference is to burn close to the same amount of
fuel and come home a little less than 6% early, a proper explanation of
the time differential. considering favorable winds, True/Indicated airspeed.

Whereas the initial extra fuel explains the ability to fly most of the
extra 480 miles on the same fuel, it doesn't explain the time necessary
to go that distance. That is where the tailwinds come in. Remember that
on the planned flight we had our headwinds in Africa, where on this flight
we were still to get them off the Mexican coast. The fact that we've had
tailwinds essentially all the way explains the time necessary to go the extra
480 miles and also the fuel necessary to go the increment of 351 miles to
480 . It all works out. An engineering analysis is done at the end that
works out the fuel and distances precisely.

*A precise engineering analysis shows the 311.5# extra fuel only gained us
 68.39# of equivalent fuel at the end, 350.8 N.Mi at 5.13 N.Mi/# since we had
 to haul the extra fuel and 83 extra pounds cabin weight around the world.

Since both the planning flight and the actual flight were computer calculated using the same fuel performance curve, the comparison should be
legitimate and free of the computational errors that would certainly creep
into the thousands of calculations that would have been necessary if both
the plan and the log had been done by hand.

From the time planning started,75°W had been the final decision point on
the trip where it would be decided to come home straight to Mojave if the
Mexican and Southwestern United States winds permitted it,or alternately
to do a dogleg and come over Mexico, then up the Mexican coast. Anticipating winter winds that required a dogleg flight, the plan had been set
to fly Mid Caribbean, 15°N,75°W, direct across Mexico to Mazatlan on the
Mexican Pacific coast. That required flying over Mexico, the Mexican land
mass and the high mountains inland from Mazatlan but left a relatively short
Northwestward flight up the coast.

No one wanted to fly over the turbulence of the land mass and over the high
mountains and the decision was made to fly across Costa Rica and then up
the coast.

With the strong headwinds, right on the nose after the Bay of Tehuanepec, that
were encountered and the long trip up the entire coast from Central America,
in hindsight, it probably would have been a better, more efficient flight
to do it as it was originally planned, since the winds probably would have
been less on the nose and for a shorter period of time.

However, this route would have produced considerably less miles and we did
not want to fly shorter miles, we wanted to fly substantially farther than
a great circle route, so that there would be no question of Voyager's ability
to fly more than around the earth with or without wind. As a result therefore, the way we did it was just fine.

The adventure and challenge continued almost to the very end. We crossed
the Northern valley of Costa Rica, along the unfriendly border of Nicaragua,
strong headwinds were encountered, starting at the Bay of Tehuanepec, one
of the two fuel pumps failed, which made it very difficult to get fuel from
the right side of the airplane. The engine was continually trying to stop
from fuel starvation, and it finally did. With the lines sucked dry, the
Voyager was a glider for four minutes. They had to replumb the lines,
raw gasoline, in a cockpit full of electronics. Jeana had earlier crawled
under the cramped instrument panel and changed out an autopilot gyro. The
people who worried about her making it, didn't understand that Hannibal
would have been lucky to have her in charge. She's sweet, smart and able.

Voyager appeared high, above a cloud bank to the South on a brilliant
Tuesday morning to 40,000 cheering people. They and the people of the
world seemed to realize that a major technical achievement and one of the
all time classic adventure thrillers was home. The story had it all,
smarts, guts, creativity, dedication, perseverance, everything that had
made America great was still working. It had been done, not by Government
or Corporations, but by very able people, acting on their own, on a grass
roots support basis. Mom and Pop's T shirt counter and thousands of $100
VIP contributions had provided a mission that would have cost NASA millions.
It was an accomplishment that Americans, indeed the human race can take
pride in. Dick, Jeana, Burt, Bruce Evans, the originals deserved it all.

Our planning proved to be amazingly accurate ! But at the time that we had to make the crucial final route decisions, there was simply no way that we could be positive! Our gut feel had to be better than the calculated numbers! If we blew it all the world would know! With the gross complexity of a nine day running calculation and little sleep it was a wild experience !

THE LAST LEG

A MATTER OF JUDGMENT

DRAMA IN MISSION CONTROL

This is the story, behind the story, that was never told. The world knew the drama in the sky. What the world didn't know was the other drama unfolding in the Performance Group in Mission Control -- Because there are times in life when the right move is to just do your job and keep your mouth shut -- and this was such a time.

Way back before Africa, Performance had made the basic and ultimately very correct judgment that the fuel burn data coming from the plane would ultimately prove to be WRONG, that it would be biased in the optimistic direction by a few percent, since Dick would be reporting when things were under control and there were simply too many times when things were pretty tough in the cockpit, and the average was not likely to meet the reports by a significant amount.

That said that no matter how well we did the flight calculations by computer, calculator, or hand, that we would get a WRONG answer, too optimistic, if we believed the numbers, because the data would prove to be wrong. The error might be only 2 or 3%, optimistic, but $2\frac{1}{2}$% of the ball park fuel expected to be used, that loaded minus the planned and hoped for reserve (7,011.5# - 400), was an error of roughly 175 pounds of fuel! That is a significant error compared to the fuel that would be left, the planned reserved, 571 pounds or the real 400 pound target, but a huge and unacceptable error if things were really close at the end. (There was finally only 106 pounds, and the potential optimistic error was bigger than the amount actually left!)

THAT MEANT THAT THE FINAL AND MOST IMPORTANT DECISIONS OF THE FLIGHT HAD TO BE MADE ON JUDGMENT, NOT A FACTUAL CALCULATION. You must dutifully do the calculations and thoughtfully consider them and their potential error, but you must ultimately make a judgment call based on the prior planning, the calculations and your gut feel for how things were really going, the underperformance or overperformance of the plane vs the basic flight and fuel plan. You would finally be forced to deal specifically with a problem you only thought was there. Worse yet, you could only guess at how big it might be!

Jack Norris knew that at some point, most likely the Caribbean decision point, he was going to have to make a gut guess of how things really were, were they going to make it or not, how much margin was there really, as the final routing decisions were made. If the routing decisions were wrong, it would be his job to change them!--And he would be semi blind!

Late in the flight, as the tanks ran dry, we would have an accurate check of the fuel left (and we did, but we also found 109 pounds had drained from the leaking left tip fuel cap), but the semi blind gut guess, when the final routing decisions were made, would be the crucial decisions of the flight from a performance standpoint, and they would be made before the final accurate answers were known.

Worse still, we didn't have all the fuel burn data on the ground, none covering the climb portions and as you've seen understanding all the variations from plan can be a real mind boggler. There was a genuinely hairy problem here that had to be dealt with intelligently or we could blow it on the world stage!

Mike Melvill noted and commented two or three times that Jack Norris was always there. Jack had quietly made a conscious decision that he had to be there 20 hours a day, so that he had an accurate continuing feel for what was going on, so that the final gut guess he knew was coming, could be right. There were risks in that, because lack of sleep impairs one's judgment and ability to do calculations, and he recognized that clearly. (Gil and Isabel Fortune, conservative and with good mature judgment, not realizing what he was really doing, kept telling him to go to bed. Normally that would have been the right advice.) Jack had sharp young guys actually doing the calculations, and they could be trusted, so he elected to let them do the math and trade his sleep for feel. Only he had done the six months work to know how all the numbers worked out, so he had to make the final quiet judgment. No one else realized the complexity!

His gut judgment at the Caribbean decision point was ultimately that they were about on Plan -- which by great good fortune proved to be more accurate than he could have possibly reasonably hoped, only three gallons off. On Plan meant that we had plenty of fuel -- if we flew the plane frugally -- but we would not!

If we got home with the 571 pounds planned reserve, we would have roughly 3,500 Nm extra range at 6 Nm/Lb. If we got home with the actual target of 400, which gave us allowance for expected extra miles and climbs, we would still have 2,400 Nm range, a transcontinental margin.

The sound judgment that we were doing as good as plan to that point was made by recognizing that we had had no African headwinds, and had had strong favorable winds in the Atlantic, and all this would make up for any reasonable performance shortfall of the plane due to exceptionally slow flight and adverse weather and the expected extra miles, and climbs. The extra fuel loaded would also help with a few hundred extra miles. We would still face the Mexican headwinds though, so we would actually have less margin than the basic plan ultimately.

Though we were arriving very early, more than 10 hours at the Caribbean, this was not a big plus, because with favorable winds, we should be early. In addition, flying high, the plane would have a true airspeed gain, would arrive sooner, but would use fuel at a comparable rate vs distance, actually even more, for no net gain from that factor. We knew how to correctly account for the altitude factor only because we had correctly worked out all the flight mechanics relationships. It was originally thought altitude was favorable, but it is increasingly unfavorable unsupercharged, above optimum full throttle altitude.

Having made that judgment correct, the final routing and flight method decisions still had to be made. Dick was already going faster and turning higher RPM. Larry Caskey and others didn't want to fly across Mexico with headwinds and high mountains. Weather wanted to cross Costa Rica and fly up the Mexican coast facing headwinds and a much longer distance. The longer distance made the record much longer, but it would be a disaster if we misjudged and fell short. The decisions no longer had big margins. Not being able to believe the calculated numbers was now a genuine problem.

We would go extra miles on a bigger than planned dog-leg. We would have big headwinds for a long time. Dick was already going faster, turning higher RPM. Burt and Dick would decide to take it easy on the engine and run it less lean at higher RPM, down from a strangulation diet. The big margin was going to get used up--and the complexity was gross and blind!

What Jack knew, having done all the homework before, was that if they in fact had planned fuel at the Caribbean, that even with all those penalties, they would make it. If he was wrong though, and of course he could be wrong, it would soon be a public disaster. It was a time to make your decision, shut your mouth and pray that your hard work and your gut guess was right. Not knowing what was still to come was the killer!

Finally, of course, losing the 109 pounds of fuel from the leaking left tip fuel cap and coming home on two engines, the flight would end up dramatically close. What really happened was that Burt Rutan's original judgment that we needed a "good margin" was correct -- we used it. The usage was largely voluntary. We were stuck with the headwinds, but we could have and would have come home more direct, leaner and not on two engines if we were in trouble. The big problem was that it had to be a judgment call! -- and the final fact would be that the amount actually left was less than the error Jack believed was in the numbers!

The sobering point was that once we were committed to the longer route via Costa Rica, we were committed to the miles and the wind,* and if Jack was wrong, Jack would properly be the "goat". The wind could be a real problem. A 20-30 knot wind when you are supposed to be going 64 knots at the flight's end almost stops you. All that was going through Jack's head, and with three hours sleep at night, that's as much trouble as a guy needs. No one else even recognized that decision was made at the time, but they would later, in hindsight, if he was wrong.

Jack knew more than anyone else, even Burt, about how it was really going, because he was the guy that did the planning, but he was purposely not rushing on to television, he was saying very little. If he was wrong, it would be a real disaster. A few in Mission Control decided he didn't know what he was doing, because he was saying little, except everything was okay. Not understanding at all, they thought it was just arithmetic. It was a calculus problem, a real mind boggler, and the numbers for the arithmetic solution were neither complete nor correct! Jack just smiled and ignored them. He was tired enough, he didn't much care what anybody

*There were even stronger headwinds forecast on a direct route home. There would not be a large penalty for the extra miles if the forecast was right.

thought. He simply had to be right. When you're that tired, you just don't care what people think -- anybody.

The main point though was a completely different point. When all the world is watching the world's greatest flight, a genuine technical wonder -- and properly impressed -- the technical guy in charge of calculations doesn't blow the image by saying he doesn't believe the data, the numbers, and he was depending on his judgment. That would improperly hurt the very proper, very favorable good image, because people would not understand all the esoteric detail. It was a time to just be quiet. The world would never understand the gross complexity of the flight and the calculations, and that the potential data error was very small but very crucial. We weren't behind the numbers, we were ahead of them. Our judgment was good enough to head off a mistake, so it would be dead wrong to hurt the image.

What Jack finally did was believe his planning work more than the actual flight calculations. The judgment that the data and thus the flight calculations would be 2 - 3% optimistic, was very fortunate, because the final accurate number was 2.69%, 178 pounds of fuel, right on. There's no sweat in a hindsight evaluation, but there was a bunch on the foresight.

There was going to be one last goal line problem. Jack was going to be dead tired for the finish when the tanks started running dry and the accurate end of flight fuel checks were possible. He was going to be too tired to calculate and even too tired to much care. He knew that his time of test had been at the Caribbean when the route and method commitment decisions were made. It was time for the rest of the team to be at bat -- and the rest of the team was first class.

Jack had the sharp crew of young guys, led by Brian Hobbs, to do the final calcs so they would be done well. Unbeknown to Mike Melvill, Jack had planned to place his final bet on Mike to be the trusted and competent friend of Dick on the goal line. Mike would be tough and smart, and reasonably rested and he could be depended upon to be there when the fat was in the fire. At the end, the flight would no longer be a calculus problem, it would be how many gallons do we have left? How fast are we burning them? How long will it take? That would be no problem for Mike, even if Mike were a little tired. The important thing would be to have Mike and his tough minded good judgment and experience there for whatever problem arose in the continuing drama. We could bet there would be more problems, more drama.

Brent Silver, a very good aeronautical engineer, would also be there with Mike. Brent was not one to get caught short. He was knowledgeable, cautious and correct. He always started pessimistic, Jack had clearly seen that, and proved to himself why something would work. He was the perfect checker. At great personal effort, he had created a full computer program to analyze the Voyager performance and then calculate the actual flight. He would be there to make sure they were going to make it.

After two hours sleep on the floor in Dick and Jeana's office, Jack was going out the door to be the color man for ABC's live coverage of the landing, with Gary Shepherd and Peter Jennings, when the engine died, the lines sucked dry due to a failed fuel pump. Near the end of the four minute silence, Mike said "For God's sake, Dick, start the front engine". Dick replied to his flying buddy, "I've been trying, Mike". The tension was thick enough to slice. They got them both running. Mike was there and in control and Jack left with trust well placed. They were going to make it. A subsequent call to Brent Silver showed Brent properly worrying and making sure, right to the end.

They replumbed the plastic fuel lines to get the necessary fuel despite a failed fuel pump. All but 106 pounds of the fuel was finally used, less than the error in the data. The calculated margins that looked fat were not as expected. What would have been done if there were even less? Had we made a risky poor decision to go as we did? No, Dick would have simply had to shut down the front engine, lean for max range, and come home on 2 Gal/Hr, not the 4.9 Gal/Hr that he was actually using. That team in the airplane was not going to quit or make any important mistakes, and neither was the Mission Control team.

This little story is not to make a big deal out of what Performance was doing, it's just to tell you the real story, how it really was. There was a lot of worry and sweat, we didn't have all the data, and we correctly felt we couldn't believe what we did have to the accuracy we needed.

Everyone in Mission Control was going flat out to do their best work. All of Snellman's weather guys were continually super in their judgments and in their hard work, and their detailed story would be equally interesting. The weather was a sweat all the way around. Dick said "the weather team saved our lives" and he meant it.

Don Rietzke and Dick Blosser's Communication guys had, in fact, kept in touch world-wide. Sometimes voices were so weak that only the trained ear of the Communication pros could catch the message, but that was how they came through. The equipment succeeded and they succeeded and the mission was accomplished. Larry Caskey had a full competent team in place. Everyone there in Mission Control were pros.

We tell you this little story so you'll know first hand the story behind the story. This is the Performance Report and that's the inside story on what happened on performance. Only those who put in the effort to read this log will know it, how it really was at the goal line. Just like Yogi Berra said, "It isn't over till it's over". We had lots of margin, but it all got used up. It was a sweat right to the bitter end. No one who was in Mission Control will ever forget the experience.

If you were interested enough to read all through this report, if you put in this much effort to find out what really happened, then you deserve to know the real story, not the Hollywood version. It was a wild pressure packed nine days, better in real life than any movie script, and of course it was finally a glorious success. Everyone was trying so hard and it was so difficult to cope with the unknowns and the challenges that the success was that much sweeter.

BRINGING ORDER OUT OF CHAOS
GETTING A HANDLE ON CALCULATING A VOYAGER FLIGHT

So far in writing a narrative understandable by a layman, we've purposely protected you from the chaos and complexity that Voyager calculations can be. Now as we switch gears and get seriously into actual calculations, we'll show you how you can make sense of the subject by understanding the Specific Range Curve.

As you've seen, the Voyager doesn't simply get so many miles per gallon. The MPG, or Specific Range, nautical miles per pound of fuel, improves as the Gross Weight decreases. An understandable way to handle this potentially confusing variation is to use the average Nm/Lb for each 1000 pound segment of the fuel, and add up all the range for all the fuel available. This will tell you either the total range or the range from any intermediate gross weight.

The Voyager can do significantly better than the minimum curve if flown precisely in good weather. If you fly higher, you go faster, but use fuel faster for a comparable range, but lose range if you go too high where the engine is less efficient. If you fly a little slower at .55 C_L or .6 C_L rather than at .5 C_L, you do better, but you lose if you fly too slow as we did in mid flight. Degree of leaning is a big swinger.

The flight becomes an infinite number of possible combinations. If the fuel reports give misleading data as we encountered on the world flight, you are faced with a real mess to sort out.

If you are the guy responsible for coping with and sorting out the confusing morass of numbers, you have to be able to get a handle on the mess some way. The way we did it was to believe that we would be close to the minimum curve, believe in the flight plan that used the minimum curve -- and then keep a feel for the possible variations from the norm. That worked. With weather impaired performance, but no headwinds yet, we were only three gallons off at the Caribbean.

Grasp this simple concept of believing the plane would come close to meeting the minimum curve. Despite misleading fuel burn data, it kept us from losing control on the World Flight. Grasp the concept NOW, because when you see the Specific Range Curve in the numerical analysis section, it becomes a busy piece of paper that is the foundation of the whole flight analysis.

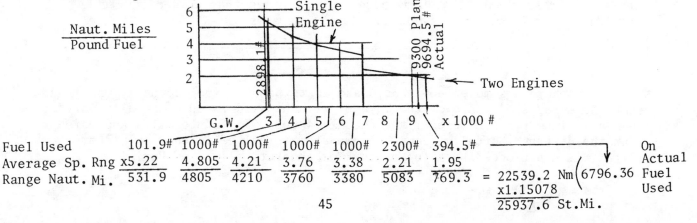

Fuel Used	101.9#	1000#	1000#	1000#	1000#	2300#	394.5#			On
Average Sp. Rng	x5.22	4.805	4.21	3.76	3.38	2.21	1.95			Actual
Range Naut. Mi.	531.9	4805	4210	3760	3380	5083	769.3	= 22539.2 Nm	6796.36	Fuel

$$x1.15078$$
$$25937.6 \text{ St.Mi.}$$

45

THE SIMPLE PERFORMANCE OVERVIEW OF THE VOYAGER FLIGHT

The refined Flight Log shows that the wind carried Voyager 2111.9 St. Miles.

The following numerical analysis shows that Voyager lost 1691 St. Miles (1469.4 Nm) on below curve performance (976 Nm was voluntary).

With the 25,937.6 St. Miles actual fuel range just calculated that equals the 26,358.6 St. Miles calculated ground miles traveled.

We gained 2111.9 St. Miles on wind, lost 1691 Statute miles on below curve performance, most of which was voluntarily given up coming home richer, faster, higher RPM and on two engines.

	Statute Mi.	Nautical Mi.
Our calculation for range was (fuel used*)	25937.66	22539.2
Our wind miles were	2111.9	1835.2
Total Available Range	28049.56	24374.4
Our lost performance was	-1691.0	-1469.4
Calculated Range	26358.56	22905.0
Actual Flight	26358.6	22905.0

So that's the simple way to look at the Flight of the Voyager. It of course took a lot of difficult work to make it that simple for you to understand. Since it was a piece of history, it was worth going through the nitty gritty detail on a one time basis to be sure we really had it right and get a genuine grasp of what really happened.

If you're really interested in Voyager and in how airplanes work, do not quit reading yet. In the next few pages, we start getting into exactly what the Voyager did -- and the narratives give a wealth of insight on how airplanes work, and how long range flight works.

The following numerical analysis does not get into the ridiculously detailed cross checks that were necessary to prove that we had it right on this piece of history, until the last few pages.

*Note that there was more fuel aboard and lost than the fuel actually used and that the unrefueled range of the Voyager without wind is substantially farther than a maximum great circle of the earth, 24,903.095 statute miles and more than the distance actually flown. Flown at its real efficiency on a summer flight, it can proceed back to New York or theoretically even Paris!

A POUND OF FUEL IS NOT A POUND OF FUEL

Pilots use simple arithmetic to calculate their flights. They are in for a rude shock if they look close at the flight of the Voyager. It is in fact an amazingly complex ever changing intricately interrelated calculus problem done by thousands of arithmetic calculations. One of the hard things to get through your head is that a pound of fuel is not a constant but an ever changing variable commodity worth only 2 Nautical miles heavy at the beginning of the flight when power requirement is high, but worth up to 6 or more Nautical miles light at the end of the flight when the power requirement is very low.

It would be relatively simple if the new thinking stopped there -- but it does not! Comparisons get complex in the extreme!

If you put in 10 pounds of fuel at the beginning to have more reserve at the end, you do not end up with 10 pounds at the end, but rather (28.4%) 2.84 pounds, because most of the 10 pounds is used carrying itself around the world. Notice that this argues persuasively that it is not twice as hard to double the record, but 3.52 times as hard. You get a good bit less than 1 pound yield for each 3 pounds of fuel that you add, only .852#.

Thus, when we put in 83# of extra load in the cabin over the planning weight and 311.5# of extra fuel, we only gained 68# of fuel at the end, a 21% yield, because we had to carry both, the 83# and the fuel!!! All this still seems reasonable, but of course you can see the mathematics getting much, much harder.

The real mind boggler is the converse of all this. If you burn 100 pounds of fuel climbing in mid flight as we did -- it does not cost you 100 pounds -- because you have now lightened the plane and it now flies on less fuel. The engineering analysis done at the end of the report shows that the mid flight climbs cost 100 pounds at the time, but really only cost 73 pounds of fuel at the end!!!! You burn fuel and it is physically gone, but you find at the end of the flight that you did not really lose that much. The physics seems to manufacture fuel and put it back in the tank. Of course all that happened is that the plane was subsequently lighter and burned less fuel, so the net effect was to lose less than you had calculated at mid flight.

If you grasp that, you have gained your first real insight into the fact that calculating the Voyager is a problem that will have you for lunch if you naively approach it with any significant degree of dumb. Fortunately Performance had been all through this by the time of the World Flight and understood how really complex a problem it was, avoided getting sucked in to making calculations that would be wrong.

A fortuitous thing happened when the crew added 311.5 pounds of extra fuel, along with the 83 pounds of extra cabin weight. Normally expected extra miles, climbs and inefficiencies would have to come out of the planned 571 pound "end of flight reserve" for the planned point to point route. The 311.5 pounds extra was just 18 pounds more than the 293.2 pounds extra actually burned by the 75°W Caribbean decision point. That made possible a straightforward comparison between the plan and the actual flight --- free of all the obscure complexities above. Thus a simple comparison is both possible and reasonably valid.

The real truth, however, is that what happened was really much more complex than that. In simple terms, most of the 293.2 pounds extra required was the cost of carrying the extra fuel and extra cabin weight around.* The lesser but still significant part was the cost to that point of the extra miles to that point, the cost of mid flight climbs, the cost of below curve performance to that point, all balanced against the help of extra winds to that point, which were substantial, because we had not hit African headwinds, but were still to hit our Mexican headwinds.

As shown by the end of flight justification in the Engineering Section, those who hang in to the bitter end will find the end result of the complexity demonstrated above is that it is necessary to convert any fuel used to equivalent fuel at the end of the flight to justify and compare all fuel costs. The people who get that far will have passed the graduate course.

*At 75°W, 233.2# of the 293.2# cost to that point was the cost of flying 311.5# of extra fuel and 83# of extra cabin weight. 60# was the net cost of extra distance, the lack of headwind to that point and all below curve performance loss. (Do you see how totally wrong it would be to compare 293.2# with the 571# planned reserve! If the extra fuel and cabin weight had not been there, the 233.2# cost of hauling it wouldn't be there!)

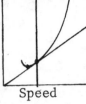

Drag #

Total

Body

Lift

IAS

A CONCISE EXPLANATION OF HOW LONG RANGE FLIGHT WORKS

The drag of bodies increases as the square of the speed, twice the speed, four times the drag, but that is misleadingly incomplete on a flying airplane, because there is also drag due to lift and lift drag <u>decreases</u> as speed squared --- $1/V^2$

Slow, a wing must fly at high angle of attack at high drag, to lift the weight, whereas fast, it flies flat at a low angle of attack, at a low drag coefficient. The rapidly decreasing drag coefficient has a greater effect than the normal V^2 effect, so between slow flight and normal cruise the drag due to lift decreases rapidly, (especially slow at first).

HP

Speed

At the speed where the two drag curves cross and are equal, the combined total drag is least. At that indicated airspeed for a given weight you find the minimum drag, the optimum angle of attack, the max lift to drag ratio, and the best speed to power ratio. (<u>Not</u> the minimum power)

The angle of attack is the key that holds the optimum relationship of lift, lift drag and body drag for the max L/D ratio, the most efficient flight.

As the weight increases, to hold that optimum angle of attack, the indicated airspeed must increase, actually as the "square root" of the gross weight ratio (simply, the converse of lift, thus weight, being proportional to the "square of the speed"). <u>Thus, there is an ideal speed curve that relates the IAS to the square root of the gross weight</u>.

IAS

GW

Flying at a constant angle of attack, a constant L/D ratio, drag goes up exactly in proportion to lift and weight as the speed increases, an easy to understand simplification.

If altitude is changed, both the proper IAS and the drag stay exactly the same for a given gross weight, because if indicated airspeed stays the same, the dynamic pressure (at the end of the pitot tube) that the airplane sees stays exactly the same. The true airspeed goes up to exactly counter the air density drop. Actually TAS increases as the square root of the air density ratio decreases.

Range vs altitude is very interesting. At a given gross weight, if you go up in altitude but hold IAS constant, drag stays exactly the same, but since true airspeed increases and power is drag times speed, as speed increases at constant drag, power goes up, as does fuel consumption, exactly in proportion to speed, and <u>range tends to stay exactly the same</u>. Actually range decreases a small amount because available manifold pressure and engine efficiency drop.

Mi/#

GW

As weight increases as above, drag goes up in proportion to weight and as above, speed goes up as the square root of weight -- Now since power is drag times speed, power goes up as $Wt^{3/2}$ and fuel comsumption, Gals/Hr, goes up as $Wt^{3/2}$. MPG or N.Mi/# is Miles/Hr / Gal/Hr which is the $\sqrt{W}/W^{3/2}$ which is W^{-1}, the inverse of weight. If weight doubles MPG drops to 1/2.

Long range aircraft fly on two curves. Speed, IAS, proportional to $\sqrt{GW}$, and the Specific Range Curve N.Mi/#,(close to) inversely proportional to GW.

HOW GOES IT CHART

On a long flight, military pilots often make a "How Goes It Chart", which shows the planned fuel usage and fuel remaining at each check point. This makes a handy and manageable reference in flight to make sure that the flight is proceeding successfully with enough fuel.

The planning "How Goes It" for the Voyager was made using the longitudinal lines around the earth. If Dick arrived at a certain longitude above the line, he knew he was ahead of plan, had enough fuel, independent of winds or extra miles traveled.

After the flight, a slightly different type of "How Goes It" is most logical. Since we didn't have actual check points exactly on the longitudinal lines, extra calculations would be necessary to show the actual fuel used and remaining at those exact points. This would produce additional chances for error and significant extra work.

Since in evaluating the flight log, we had calculations which showed how we were doing at each major checkpoint -- and had carefully corrected the comparison, so there was a valid comparison where the checkpoint plan and the actual checkpoint were at a slightly different latitude and longitude, it is logical and proper to use these actual points in the after flight "How Goes It" plot. One additional difference sneaks into this method, however. Since the actual miles traveled at each checkpoint is always longer than the planned miles at each checkpoint, comparable checkpoints plot at slightly different distances in each case.

This is really not a handicap, however, once one understands what happens. If you're doing well at a checkpoint, the point will plot above the intended point for fuel remaining and yet at a greater number of miles distance. A plot made in this way allows you to see when the fuel is good, that is, above the original checkpoint -- and also shows you how many extra ground miles you actually traveled.

As a result, the following post flight "How Goes It Chart" needs to be read slightly differently than a normal military "How Goes It Chart", but in fact, gives one more information, the extra miles traveled, in addition to how well you're doing on fuel vs the plan. As above, this is simply seen by how far the actual point is above the planned fuel remaining.

The clear message of the chart is that the fuel disappeared at the very end -- when we were flying on 2 engines at 4.9 GPH rather than 2 GPH - against a headwind - faster at higher RPM - not leaned as much as normal.

50

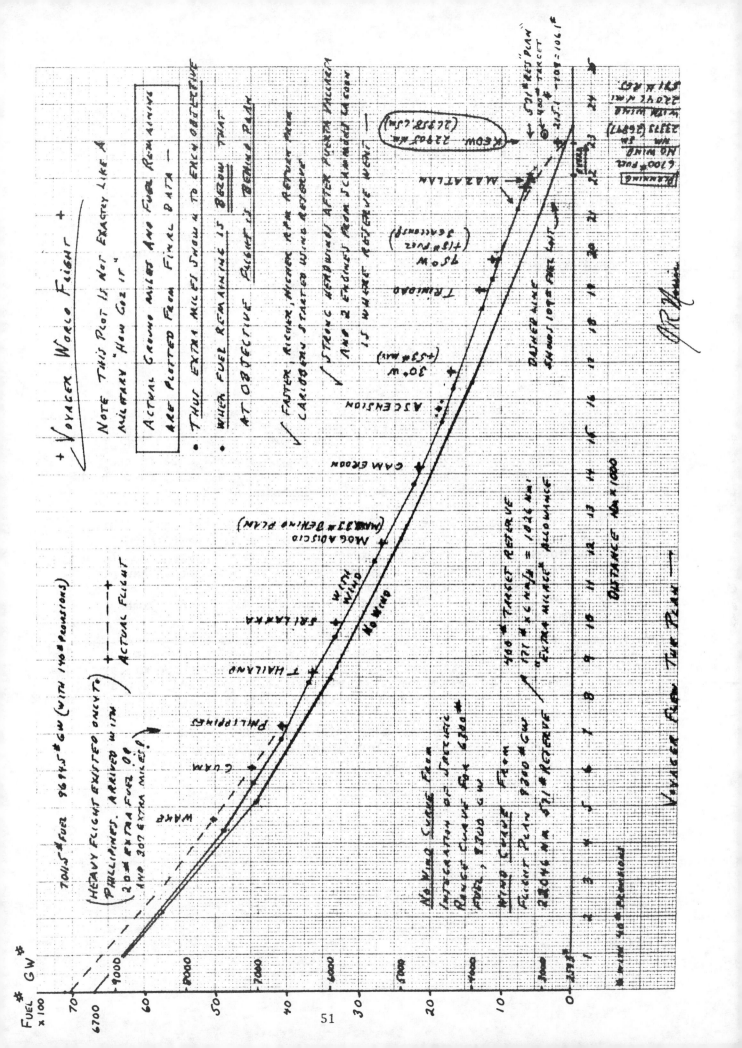

THE SPECIFIC RANGE CURVE FOR THE WORLD FLIGHT

The World Flight Specific Range plot is a very busy piece of paper but it has a wealth of information.

Starting at the bottom of the page, it recalculates range for the actual World Flight for 7,011.5# of fuel, a 2,683# plane with crew and provisions, a 9,694.5# gross takeoff weight. Range is 23,693.7 N. Mi, 27,266.3 Statute Air Miles. With wind, that's a 29,000 mile range. A 30,000 mile range on a favorable summer flight could be achieved by simply getting .15 N.Mi/# above the minimum curve. Clearly there was excellent margin and with a good crew, success was assured, unless we encountered a mechanical, or weather problem large enough to stop the flight.

Next, the actual air miles at each 1000# GW point was added from the re-fined computer calculated log, and the total air miles at the end, and the calculated air miles range at the calculated landing weight was added.

Note, that based on fuel used, the calculated lower curve range was 22,539 N.Mi, and the actual flight for the fuel used was 21,069.8, a 1,469 N.Mi perfor-mance short fall. Notice however, that at 4,000# GW, just before Trinidad, we had only lost a net of 533 N. Mi for the flight, very little in the overall. We voluntarily gave up 976 N. Mi by coming home rich, fast, high RPM and finally on two engines, a voluntary decision at the end.

The plot of the specific range and the speed vs GW tells the whole story. We lost most of the 533 mile range in weather climbs and the rest in mid flight losses, first trying to get up to the curve in weather, high, just after front engine shutdown, and by being too slow as a result of having an erroneous gross weight prior to Somalia. Notice that at 7,000# GW we were 153 N. Mi ahead and thus had lost 686 miles in mid flight climbs and short falls, then gained 40 N.Mi back by 75°W.

> The specific engineering analyses that follow at the rear of this report are based on the data on this page, by specifically analyzing the mileage and fuel usage by flying the minimum curve for the actual, and the planned flight and by comparing those with the actual results.

A separate analysis of climbs was made which total 100# of fuel in mid flight, 56# for the climbs and 44# for "maintaining the higher powers required at altitude" for the time involved.

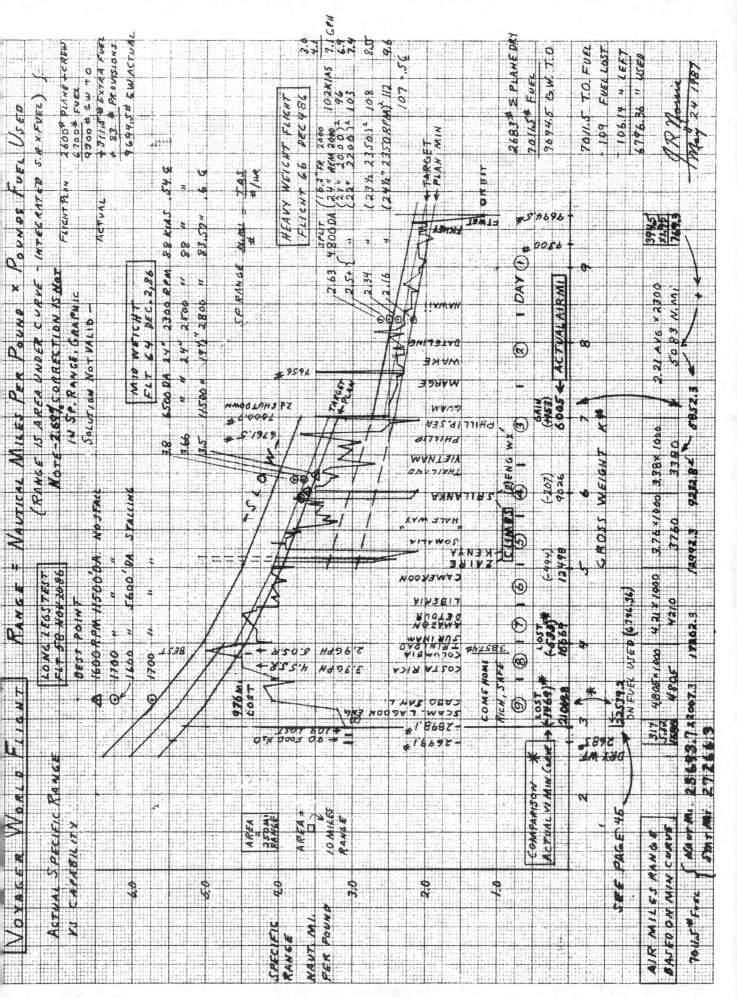

VOYAGER WORLD FLIGHT
RANGE ≡ NAUTICAL MILES PER POUND × POUNDS FUEL USED

SPECIFIC RANGE

NAUTICAL MILES PER POUND OF FUEL

Burt Rutan had created from all the early flight test data, a specific range plot showing the minimum and best fuel performance vs range in nautical miles per pound that would be expected for both single engine and two engine operation over the entire gross weight range.

The performance was tested throughout the entire summer of 1986 through the fall, up until Dec. 2, twelve days before World Flight liftoff. Based on testing, later more sophisticated versions of the curves were produced but the original data was basically excellent and in the interest of consistency and ease of understanding the same upper and lower curves were used right up through World Flight. Actual testing showed that the minimum curve was hardest to meet just after front engine shut down, but with aggressive leaning the minimum curve could readily be met. Testing also showed that even better performance was attainable light. For reference key data conclusions were actually plotted on the final curves used, but the upper and lower curves were purposely not changed.

All planning was based on the minimum curve and the simple objective was to actually try to meet that minimum curve under the much more difficult conditions that would be encountered on the actual World Flight compared to the test flight conditions.

The only basic change to the original curves was as below to show that the range was basically independent of altitude but slightly degraded at altitude because of lower available manifold pressure which slightly lowered engine efficiency.

Aggressiveness of Leaning

With max power obtained at about a 13:1 air fuel mixture on an engine with perfect fuel distribution and max economy at a 16:1 air fuel mixture, there is a very big, 16/13, 23% difference in fuel economy depending on how aggressively an engine is leaned.

This factor is so strong, so important, that it becomes the overpowering driver in fuel economy testing and it is often not possible to distinguish the effect of more subtle factors because with small differences in the aggressiveness of leaning, the data on other factors is swamped and looks inconsistent.

Other factors such as days with variable turbulence and possible aerodynamic variables, even possible differences in fuel induction and distribution made interpretation of fuel economy test data difficult at best. Thus it was that after months of testing there was a limited number of test days where one felt that the data should be trusted on the more subtle points.

There was no doubt whatsoever that aggressiveness of leaning was the chief factor.

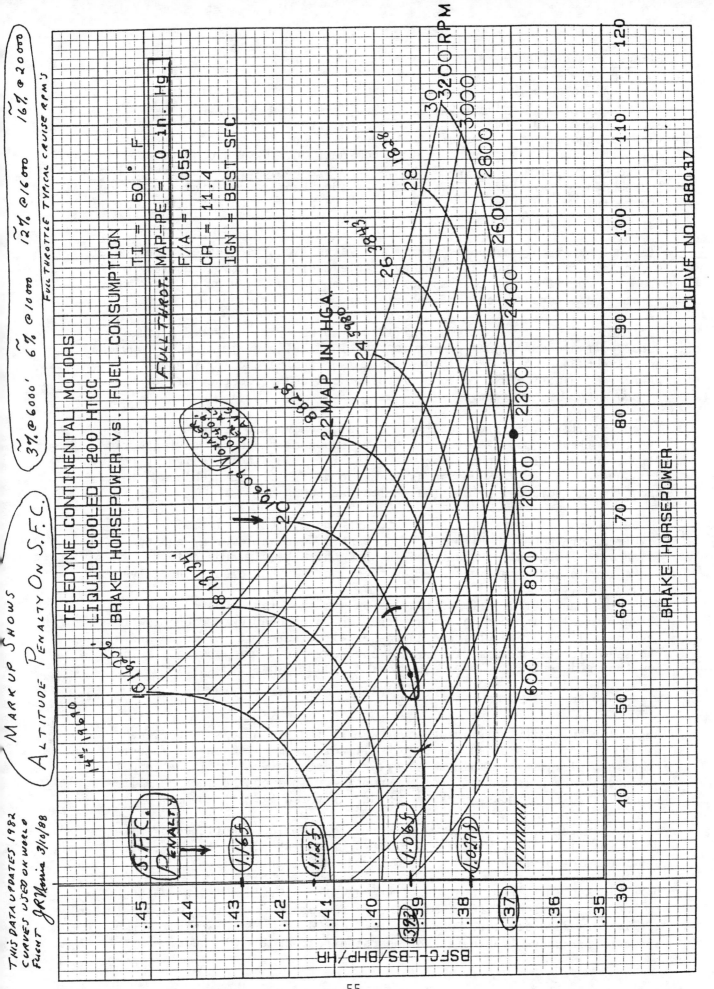

Effect of Altitude

For each gross weight there is theoretically an optimum Indicated Air Speed and _if that IAS is held precisely_, drag is the same at any altitude. True Air Speed, and thus power increases as the (square root of the) density ratio of the air decreases with altitude. If the engine and propeller efficiency would stay the same, fuel comsumption, power and speed would all go up by exactly the same amount and range would stay exactly the same, independent of altitude. This is a most important fundamental conclusion. You go faster and burn fuel faster if you go higher, thus you tend to go as far but no farther at altitude.

If you look at engine curves, there is a significant*decrease in engine efficiency as altitude increases and available manifold pressure decreases. The best specific fuel consumption goes from below .4#/HPhr to above that as altitude increases on the crucial rear engine for example.

Testing showed that it was most difficult to meet the targeted minimum specific range curve just after front engine shutdown, just below 7000# GW and flight 64 on Dec. 12 at 6300# was a late and particularly good test where the objective was to do everything well and prove the ability to be above the minimum curve.

Flight 64 with aggressive leaning was indeed above the minimum curve and its fuel final composite data points are shown on the accompanying World Flight curves. These points show the specific range to be slightly better at 6500 ft than at 11,500 just as the best most believed previous data shown in the adjacent points (unlabeled) at 6000# GW.

To be precise it must be pointed out that the altitude point of course required more power, a higher RPM and was flown at a lower C_L and on balance review of all the data would show that the altitude effect was a little less than indicated by these points, but clearly negative.

The Performance Group's final conclusion was that range was as theory shows basically independent of altitude but biased a few percent lower as available manifold pressure decreases with increasing altitude and the specific fuel consumption of the engine rises, a few percent over a few thousand feet.* A partially closed throttle hurts efficiency low.

Effect of RPM

Logically an engine turning at lower RPM is fed its necessary correct air fuel ratio less often and though all aerodynamic and engine factors come into play, in general the lower RPM 's at a given condition tend to produce the best range and fuel consumption. This can be seen on all the selected data shown on the accompanying World Flight curve.

The good catch phrase put forward by the ferry pilot experience of Mike Hance was "The slower it turns, the less it burns" a correct principle shown earlier by Lindbergh in World War II and in his writings.

*6% at 10,600 feet, the Voyager average density altitude, 16% at 20,000 feet! If you fly a narrow altitude band only a few percent are involved, and it can get lost in data scatter. Poor mixture distribution of throttled engines prevents agressive leaning and hides the effect at low altitude.

A most interesting extreme case was shown on the very low RPM "long legs test" Flight 58, Nov. 20, 1986, where very high, indeed extreme propeller angles of attack were forced by very low RPM. At 5600' density altitude at 1600 RPM with the blades actually stalling inboard, better data was obtained at 1600 RPM than at 1700 RPM where the stalling and shaking subsided!!! I would never believe that until I saw it myself.

Also worthy of note, at 11,500, where more pitch angle for more speed at the same IAS but a higher TAS was needed and it was possible to fly at 1600 RPM without stalling, the higher altitude data beat the lower altitude range data! This was the only place in all the test where higher altitude range beat lower altitude range!

These very low RPM tests produce spectacular range numbers, up to 6 3/4 N. Mi/Lb at less than 2 GPH and this at above 3000# GW, actually 3250, well above the final flight weights possible.

Near the end of the flight, with some expected lead fouling after 7 days, the decision was made to come home richer and faster at higher RPM and then finally on two engines, and these suberb range conditions were never used on the World Flight. Dick's lean was right on the edge of strangulation!

Best Coefficient of Lift

The entire flight test program was aimed at the initially planned .5 C_L up until flight 64, Dec. 2, 1986, twelve days before the Flight. A .5 C_L speed was known to be above max L/D but was initially thought to be necessary for improved controllability.

As usual the data had scatter and was not easy to interpret, but it appeared to Performance that .55 C_L rather than higher C_L's was the best. Some .6 C_L data looked good and in a heavy tailwind condition there was a clear argument for a slower, higher C_L for max range.

Fan and turboprop aircraft recommended cruise is always set significantly above max L/D speeds since performance loss is small and there is a substantial speed gain and recognizing this, Performance recommended a .55 C_L World Flight as actually more efficient, faster, and essentially equivalent even with some wind.

The problem was that there was no time for intelligent discussion of the subject in the crunch that developed and even with somewhat difficult to interpret data, the hard fact was that anywhere in the .55 to .6 C_L range results would be good, though the flight would be 3 to 5 knots slower at low and high gross weights respectively if .6 C_L was used.

Performance could have been downright angry at the last minute decision to fly slower and ultimately on a completely undefined speed curve because it blew four months planning right out of the tub and with no warning, but in fact after a lifetime of creating products and knowing how the minds of creative people work, such switches came as no surprise. With a foot high stack of 15 computer runs and a wild three days before launch, we were again under control and ready for launch by mid afternoon the day before.

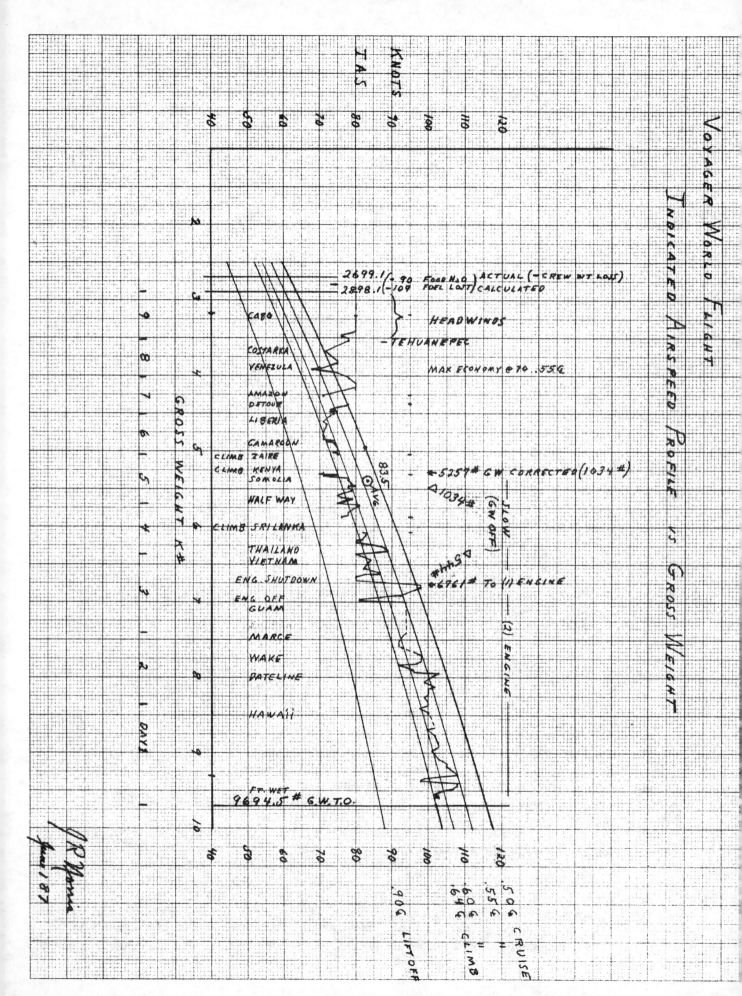

VOYAGER WORLD FLIGHT — INDICATED AIRSPEED PROFILE vs GROSS WEIGHT

WORLD FLIGHT

ANALYSIS OF C_L AND SPEED

RELATED TO SPECIFIC RANGE, NAUTICAL MILES PER POUND

The Speed and C_L of the World Flight   (Ref Actual Speed vs GW Plot Following -
                                             and Composite Speed and Specific Range Curve)

On any 9 day plus flight, crew fatigue could be predicted to produce a
flight profile that varied from perfection. People are not built to follow
a little needle for 216 continuous hours, day and night, with little sleep.
When the decision was made to go for a December winter flight with con-
siderable weather interference, it was clear in advance to Performance that
the flight's adventure, difficulty and risk content would go way up, and
the performance and precision of the flight would without question suffer
to some degree. The data plot of the actual result looks quite erratic
until you realize we are only talking about a few knots from the ideal,
and that quite often the game was not precision but rather surviving.

For the first few days, except for some experimenting with limits and
some minor wandering from ideal, the speed vs GW plot shows the Voyager
stayed generally in the 5 knot range bounded by .55 C_L and .6 C_L, as
Dick alternately flew, sought optimums and had periods where he was
either bored or busy with other challenges. Here Performance generally
met that recommended and predicted on the flight data given to the crew
before flight, falling near the specific range curve halfway between the
min and max curve which flight test data suggested would occur.

After front engine shutdown, until the increasingly wrong gross weight
was corrected as the Voyager went over Somali, the Voyager flew substantially too slow thinking the craft was lighter than it was and the actual
corrected specific range data showed the Voyager underperformed the minimum curve. As shown, it was often actually below the .64 C_L climb speed
curve, far too slow for best results due to the gross weight discrepancy.

The negative bias on speed just before the 1000 pound plus GW correction
was made over Somali would be 7 knots as seen by the curve for 1000 pound
difference, so average speeds were closer to a .64 C_L climb speed rather
than between .55 and .6 C_L where it should have been.

Once it was realized the gross weight was heavier and was corrected over
Somali, the Voyager generally speeded up and generally began to meet the
fuel performance curve. Though that is true, to be accurate it must imme-
diately be pointed out that a close look at the actual plots of IAS and
Specific Range vs GW showed the facts masked by climb periods (when
no actual speed or fuel burn data was reported to us, but rather had to
be calculated) and weather periods where Voyager was still too slow
fighting other problems. and properly slowing up in tailwinds.

As shown later in specific numerical analysis,with climbs and slow and below minimum range performance, Voyager lost 689 miles of range in this too slow middle gross weight period. Hundreds of miles of range loss, however, was <u>not</u> a problem because Voyager had much more than <u>hundreds</u> of miles extra range. Until the end, wind masked the shortfall.

Though the Voyager was $4\frac{1}{4}$ hours late reaching Africa, it had traveled 448 miles extra by Mogadiscio, and its timing had not been badly hurt despite its too slow speed, simply because it had been flying higher on average than the standard plan. TAS increases countered IAS decreases.

Higher peaks in specific range shown on the plot just after the climbs are invalid data shown as dashed lines since altitude was not held! It's possible that Dick ran richer in the climbs struggling to climb over the tops, but since only 55 pounds extra was burned in the calculated climbs and 45 in maintaining for 100 pound total, any error would be in the range of a gallon or two, a miniscule amount in the overall flight. In the same vein, calculations were done on flight log times rather than barograph data, so that the refined log is not a hodgepodge of minutia, but rather a valid representation of the historic log. It is possible to show many small differences, but they average and are small.

When tailwinds rather than headwinds were encountered over Africa and the Eastern Atlantic and high flight over weather was necessary, Voyager rapidly went from 4 1/4 hours behind schedule to 7 3/4 hours ahead by 30°W off Brazil.

Between the Mouth of the Amazon River and Trinidad, the decreasing weight and power requirements drove the required RPM below 2000 and after one leg at 1850 RPM, Dick pushed the RPM and speed up to 80 knots, well above the .5 C_L curve, well above plan and started a <u>fast return home</u>, even before the Caribbean, which put the specific range again just below the minimum curve.

Off Caracas, Venezuela, the maximum range of 5 nautical miles per pound of fuel was reached at .55 C_L when the plane was temporarily slowed, better than the 4.9 achieved at .6 C_L in the leg just before.

Off Baranquilla, Columbia, the decision to come home rich was made to conserve the rear engine, speed had already been increased and higher RPM was used and nautical miles per pound of fuel fell. When two engines were conservatively run after Scammons Lagoon, Baja, the nautical miles per pound fell far below that achievable. These decisions which used up 190 pounds of fuel, half of the expected reserve, were the key decisions of the flight and by decision were the major departure from the available performance. It was otherwise a nominal flight where we were heading inexorably for the targeted 400 pound reserve -- with only a faster return due to higher flight resulting in an earlier return than that shown on the plan.

The incisive engineering analysis that follows digs much deeper and shows very explicitly what happened at both 75°W and at the end of the flight and compares it with the planned flight and with a theoretically perfect flight. It gives real indepth insight into the obscure workings of the complexity of the Voyager's inner secrets.

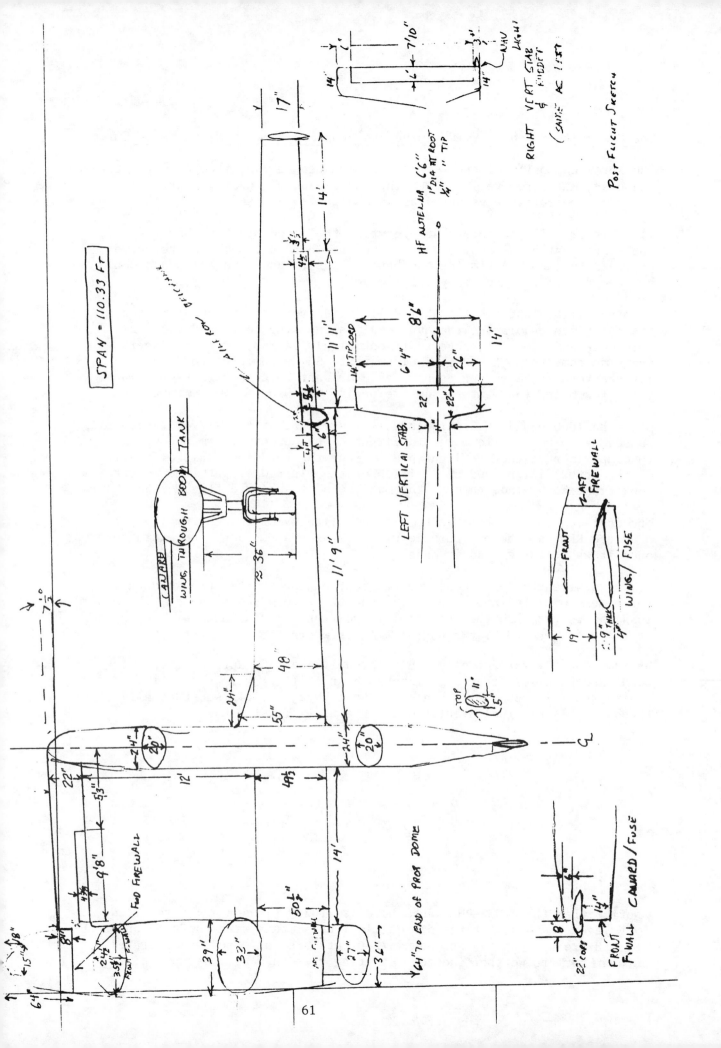

SPAN = 110.33 Ft

CANARD
WING, THROUGH BOOM TANK

LEFT VERTICAL STAB.

RIGHT VERT STAB
& RUDDER
(SAME AC LEFT)

HF ANTENNA 6'6"
1" DIA AT ROOT
¼" " TIP

POST FLIGHT SKETCH

FWD FIREWALL

END OF PROP DOME

WING / FUSE

CANARD / FUSE

FRONT
FIREWALL

The Composite Speed/C_L vs GW Curve and Specific Range N Mi/Lb vs GW Curve

The Xerox composite of the two key performance curves, though very cluttered and busy, makes for very interesting and reasonably easy comparison of fuel and range performance vs speed as GW changes.

It is interesting to note that everytime the speed hit the .55 C_L curve, the range peaked as recommended to the crew before launch on the curves provided to them. A similar, perhaps slightly less favorable case could be made for .6 on a point by point check.

Despite flight conditions that made it often more of an adventure and a test of human endurance, than an ideal engineering test, and plots of speed, C_L and specific range that look highly variable, the engineering facts do come through and it's clear that .55 C_L to .6 is the place to fly the Voyager depending on how fast you want to come home vs how many extra few miles you want to squeeze out of the wind.

Only the 1000 pound GW discrepancy in mid flight with its 7 knot slow bias caused us to underperform in mid flight. This underperformance was masked in comparison to the preflight plan, since the headwinds came off Mexico on the actual flight and over Africa on the planned flight, the more normal condition. Good winds through the Caribbean, masked mid flight excess burn.

The decision to come home fast, rich and finally on two engines, all at elevated RPM was the big purposeful decision that used up the extra reserve once it was clear we had it made.

It was a wise choice to conserve the rear engine and be sure we actually got home, completed the flight and made the records as planned. The minor loss was to not show the spectacular fuel range performance of the Voyager at the end. The 7.3 Hr two engine time is where the fuel went, at 4.93 GPH.[*]

We knew the Voyager actually could get well above 6 nautical miles per pound and burn less than 2 gallons an hour at light weight, really amazing performance. Tests had shown that Voyager could substantially exceed Burt's original prediction. Absolutely first class work.

[*] Notice that the 4.93 GPH stated here is 1.0269 times the 4.8 reported by Voyager and shown on the offical log! Consistant with the note on the log, GPH are as reported and pounds/hour etc are corrected to accurately account for the weather induced reporting bias on fuel consumption.

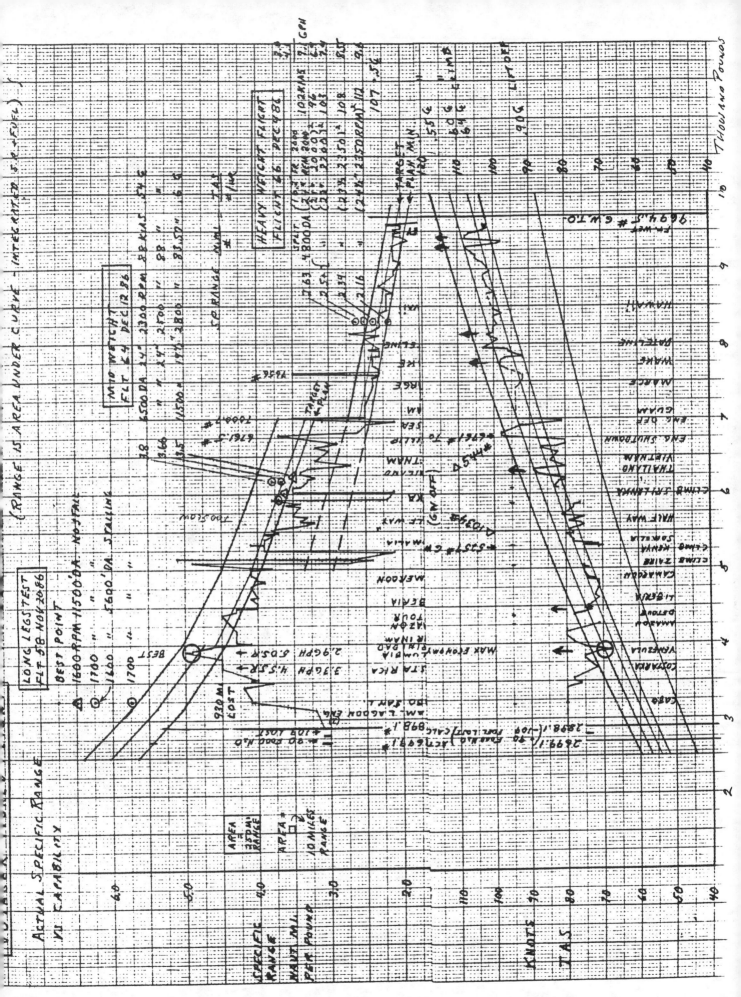

63

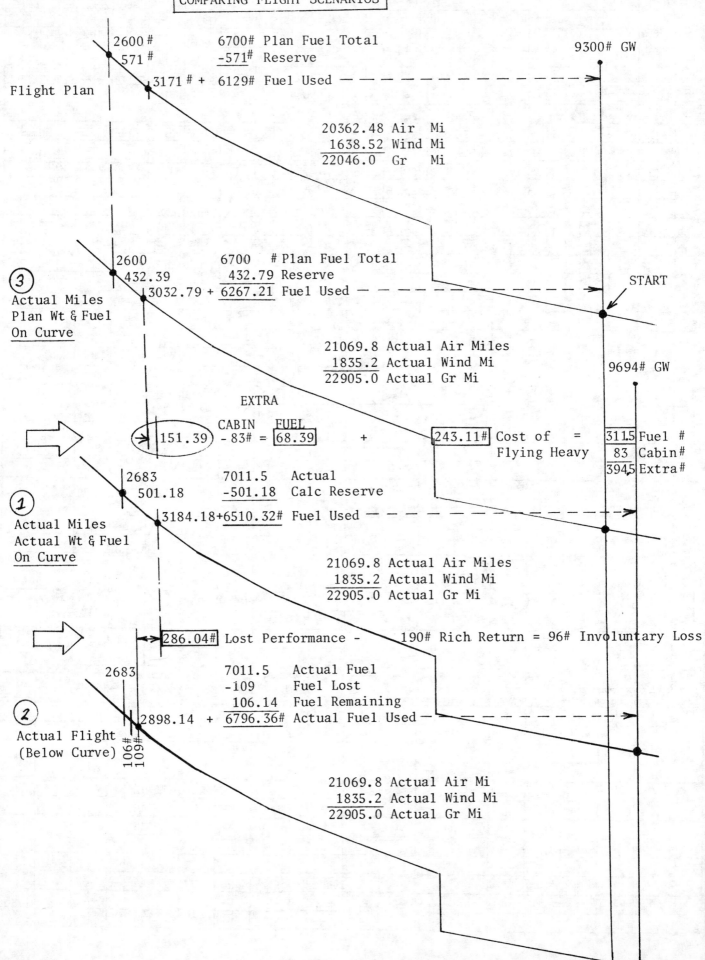

Flight Plan

2600 #
571 #

6700# Plan Fuel Total
−571# Reserve
3171 # + 6129# Fuel Used

9300# GW

20362.48 Air Mi
1638.52 Wind Mi
22046.0 Gr Mi

③ Actual Miles Plan Wt & Fuel On Curve

2600
432.39

6700 # Plan Fuel Total
432.79 Reserve
3032.79 + 6267.21 Fuel Used

START

21069.8 Actual Air Miles
1835.2 Actual Wind Mi
22905.0 Actual Gr Mi

9694# GW

EXTRA

CABIN   FUEL
151.39 − 83# = 68.39   +   243.11# Cost of = Flying Heavy

311.5 Fuel #
83 Cabin#
394.5 Extra#

① Actual Miles Actual Wt & Fuel On Curve

2683
501.18

7011.5 Actual
−501.18 Calc Reserve
3184.18 + 6510.32# Fuel Used

21069.8 Actual Air Miles
1835.2 Actual Wind Mi
22905.0 Actual Gr Mi

286.04# Lost Performance −   190# Rich Return = 96# Involuntary Loss

② Actual Flight (Below Curve)

2683

106#
109#

7011.5 Actual Fuel
−109 Fuel Lost
106.14 Fuel Remaining
2898.14 + 6796.36# Actual Fuel Used

21069.8 Actual Air Mi
1835.2 Actual Wind Mi
22905.0 Actual Gr Mi

The following analysis will provide the two keystone final check numbers that make it possible to cross check all the final fuel calculations: 286 pounds of fuel, the ultimate measure of "lost performance", and 243 pounds of fuel, the cost of flying heavier than plan.

It is appropriate to theoretically fly up the minimum performance curve from the actual 9694.5 pound takeoff gross weight for the actual 21,069.8 air miles flown and determine theoretical fuel burn required.

Actual GW 9694.5#. At 3000# GW Theoretical N.Mi = 22007.3¶   6694.5# Fuel*
                      Actual   N.Mi = -21069.8

                                         937.5 N.Mi
(3184.18# Theoretical GW at End of Flight)   ------------ -184.18#
                                             5.09 N.Mi/#

① PERFECT MINIMUM CURVE FLIGHT AT ACTUAL GW. | 6510.32# Theoretical Fuel Reqd. |

② Subtracting theoretical fuel 6510.32# from actual fuel used 6796.36# one gets | 286.04# | of fuel an exact measure of the pounds of fuel lost in voluntary and involuntary below curve performance on the world flight. This compares directly and correctly with the | 1469 N.Mi | shown to have been lost on the specific range curve comparing the actual fuel theoretical range and the actual range obtained. The proper theoretical specific range at the end of the flight is obtained by dividing these two numbers. Dividing, 5.13 N.Mi/# is the proper mean specific range between the theoretical and actual gross weights at flight end based on actual fuel usage. ⟹ (Note that the 286.04# of fuel cost should not be directly compared with the 311.5# of fuel added at the beginning of the flight which was much less valuable "early" fuel!)

Planned Flight Comparison

It is then appropriate to theoretically fly up the minimum performance curve from the planned 9300# takeoff gross weight for the actual 21069.8 Nautical air miles flown and determine the theoretical fuel required for comparison to the heavier actual flight

Planned GW 9300#. At 3000# GW Theoretical N.Mi = 21338 N.Mi 6300# Fuel
                      Actual   N.Mi -21069.8

                                         168.2 N.Mi
                                         ------------ -32.787
                                         5.13 N.Mi/#

③   (3032.79# Theoretical GW at End)(△151.39#) | 6267.21# Theoretical Fuel Reqd |

Comparison of the 6267.21# of fuel theoretically required at the planned GW of 9300# to the 6510.32# theoretically required at the actual GW of 6694.5# shows the penalty for flying at the 394.5# heavier GW was 243.11# | PENALTY HEAVY 243.11# Fuel |
(Remember 394.5# = 311.5# extra fuel + 83# in cabin)

*6694.5# fuel is the fuel used at 3000# GW (9694.5 - 3000 = 6694.5)
¶Ref Specific Range Curve at 3000# GW.

Conclusions

In addition to the two key numbers of 286.04# of fuel "lost performance" and 243.11# of fuel cost of heavier flight, two additional interesting conclusions become clear.

1. The extra 311.5# of fuel (-243.11# heavy penalty) only gave us the equivalent of 68.39# of fuel at the end of the flight 350.84 N.Mi at 5.13 N.Mi/#. Cabin weight hurts.

2. The 286.04# fuel cost of voluntary and involuntary below curve fuel performance minus the 68.39# net gain of the extra 311.5# extra fuel gives a <u>217.65 pound net deficit</u> on the actual flight <u>vs the planned flight.</u>

Explanation for Those Interested
Refer to the calculations above and the World Flight Specific Range Curve.

What happens here is that burning 394.5# more fuel (311.5# + 83#) at the heavy gross weight end only gets us 769.3 N.Mi at 1.95 N.Mi/#, a poor mileage yield, whereas we burn 151.39# less fuel at 5.08 N.Mi/# at the light end where mileage yield is much better to get the same 769.3 N.Mi. The 151.39# is the difference between the 3184.18# theoretical end gross weight starting at the actual flight gross weight and the 3032.79# theoretical end gross weight starting at the planned 9300 T.O. GW.

* <u>151.39# is equivalent to the 83# extra cabin weight and the 68.39# fuel gain, a poor gain compared to the 311.5# added at the 394.5# heavier gross weight end. You put in 311.5# of fuel at the start, but find you only gain 68.39# at the end, because extra cabin weight was also carried.</u>

This can give substantial insight into why an around the world plane must be kept light. A pound of fuel can yield 5 or 6 N. Mi at the end of the flight but less than 2 at the heavy end. Adding fuel at the heavy end does marginal good. Adding cabin weight hurts.

WITH THE EXTREMELY LIGHT WEIGHT, SUPER STRONG VOYAGER STRUCTURE WORLD FLIGHT BECAME POSSIBLE! With a normal structure it is not possible.

Though all this sounds ridiculously esoteric, the two key numbers developed in this analysis, 286# and 243# of fuel become the keystone check numbers crucial to the final check of the entire performance analysis -- <u>and you</u> don't really understand it until you see how all this works out.

THE REASON FOR DOING ALL THE FINAL CROSS CHECKING THAT FOLLOWS IS QUITE SIMPLE AND VALID. THE FINAL CROSS CHECK IS GOOD EVIDENCE THAT ALL WORK AND LOGIC IS CORRECT! OUR INITIAL PREMISE WAS THAT VOYAGER WAS A TRUE PIECE OF AVIATION HISTORY THAT DESERVED A COMPLETE AND CORRECT LOG.

THE FLIGHT OF THE VOYAGER WAS A CALCULUS PROBLEM DONE BY THOUSANDS OF ARITHMATIC CALCULATIONS. A POUND OF FUEL USED IN MID FLIGHT IS <u>NOT</u> A POUND AT THE END! IT IS A FOOLS TRAP LOOKING FOR THE UNWARY! THOSE WHO INVEST THE TIME TO UNDERSTAND WILL BE WISER FOR THE EFFORT!

COMPARISON OF POUNDS OF FUEL PLANNED RESERVE VS ACTUAL FLIGHT FUEL REMAINING

With the numbers just developed[¶] and some arithmatic gymnastics, a reasonably exact comparison of the planned flight which predicted a 571# fuel
reserve and the actual flight can be made which will give a check of the
calculation methods and numbers and additional insight on the numbers
and what they mean.

Planned Fuel Reserve   571# Reserve
Fuel required for extra miles considering extra wind.

	Act	Plan				
Gr Miles	22905 -	22046	= 859	Extra Ground N. Miles		
Wind	8.493 -	7.525	=	Tail Winds = +.968 knots		
Flt Time	216.0622	Decimal Hrs		Act Flt Time x 216.0622		
Wind Mi Gained			(-)209.15	Wind Gain 209.15 N.Mi		
Time Diff				-7.67 Hrs Wind Not Avail.		
Plan to Actual	216.06 -	223.73	=	x7.52 Planned Wind Knots		
Lost Wind Miles			(+) 57.68	57.68 Unavail Wind Mi		
Extra Fuel Reqd			707.52	Net Extra Air Mi =	-137.92#	
			5.13	N. Mi/#		

Reserve fuel if plane flew on min curve   433.08#*
Net of performance loss and extra fuel (286.04 - 68.39)[¶] =   -217.65
Calculated fuel at flight end.   215.43#
 (Note agreement with 215.1# at end of refined flight log)

Note that if 190.53# of "Rich Return" fuel were added to
215.43# of fuel the total is a 405.96 reserve, just 5.96#
more than the 400# target!!! (190.53 See next page.)

Lost Fuel - Leaking left tip tank cap   -109

Fuel Remaining Per Calculation Prediction   106.43#

Actual Fuel Remaining Measured   106.14#
                                                                  .29 Error

The remarkably good agreement gives a very good check of the
calculation methods and thinking on a comparison that is not
at first easy to see through and do correctly.

Note that net wind gain is (209.15 - 57.68 =) 151.47 N. Mi,
29.526# at 5.13 N.Mi/# which checks well with the 151.68 N.Mi
net wind gain of the following page.

The additional check calculations are done since they provide
excellent additional insight on the flight.

*Note that if we had flown this flight in summer without extra fuel, extra
climbs and met the minimum curve, our end flight reserve would be 433#,
just 33# over the target.

NUMERICAL COMPARISON OF PLANNED VS ACTUAL FLIGHT

AT 75°W AND END OF FLIGHT

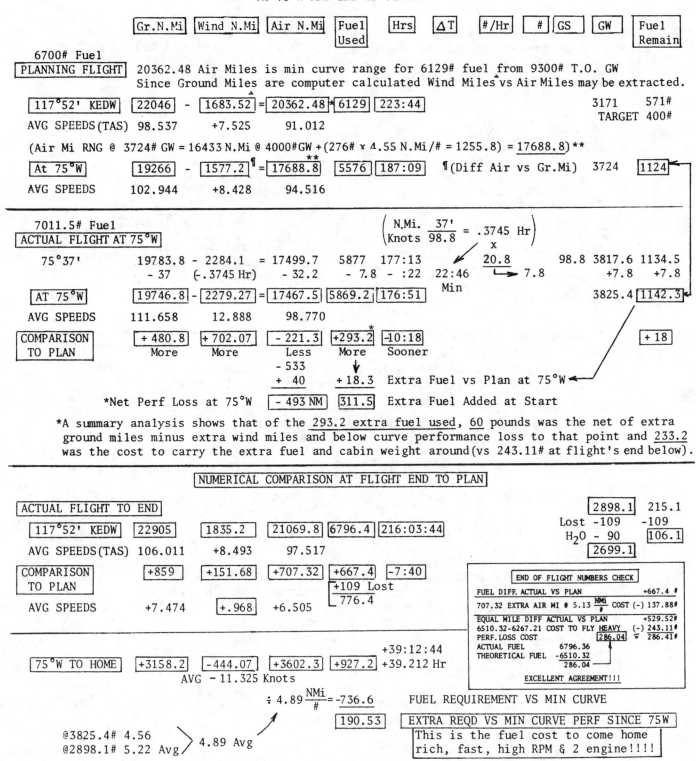

| Gr.N.Mi | Wind N.Mi | Air N.Mi | Fuel Used | Hrs | ΔT | #/Hr | # | GS | GW | Fuel Remain |

6700# Fuel

PLANNING FLIGHT   20362.48 Air Miles is min curve range for 6129# fuel from 9300# T.O. GW
Since Ground Miles are computer calculated Wind Miles vs Air Miles may be extracted.

117°52' KEDW   22046 - 1683.52 = 20362.48 6129   223:44   3171   571#
AVG SPEEDS (TAS) 98.537   +7.525   91.012   TARGET 400#

(Air Mi RNG @ 3724# GW = 16433 N.Mi @ 4000#GW +(276# x 4.55 N.Mi/# = 1255.8) = 17688.8)**

At 75°W   19266 - 1577.2¶ = 17688.8** 5576 187:09 ¶(Diff Air vs Gr.Mi) 3724 1124
AVG SPEEDS   102.944   +8.428   94.516

7011.5# Fuel

ACTUAL FLIGHT AT 75°W $\left(\dfrac{N.Mi.}{Knots} \dfrac{37'}{98.8} = .3745\ Hr \right)$
                                                              x
75°37'   19783.8 - 2284.1 = 17499.7 5877 177:13   20.8   98.8 3817.6 1134.5
            - 37 (-.3745 Hr) - 32.2   - 7.8 - :22 22:46 → 7.8   +7.8   +7.8
                                                      Min
AT 75°W   19746.8 - 2279.27 = 17467.5 5869.2 176:51   3825.4 1142.3
AVG SPEEDS   111.658   12.888   98.770

COMPARISON   + 480.8   + 702.07   - 221.3   +293.2*   -10:18   + 18
TO PLAN   More   More   Less   More   Sooner
                                   - 533
                                   + 40   + 18.3   Extra Fuel vs Plan at 75°W

         *Net Perf Loss at 75°W   - 493 NM   311.5   Extra Fuel Added at Start

*A summary analysis shows that of the 293.2 extra fuel used, 60 pounds was the net of extra
ground miles minus extra wind miles and below curve performance loss to that point and 233.2
was the cost to carry the extra fuel and cabin weight around (vs 243.11# at flight's end below).

NUMERICAL COMPARISON AT FLIGHT END TO PLAN

ACTUAL FLIGHT TO END   2898.1   215.1
                                                         Lost -109   -109
117°52' KEDW 22905   1835.2   21069.8   6796.4   216:03:44 H_2O - 90   106.1
AVG SPEEDS (TAS) 106.011   +8.493   97.517   2699.1

COMPARISON   +859   +151.68   +707.32   +667.4   -7:40
TO PLAN   +109 Lost
                                          776.4
AVG SPEEDS   +7.474   +.968   +6.505

END OF FLIGHT NUMBERS CHECK

FUEL DIFF. ACTUAL VS PLAN   +667.4 #
707.32 EXTRA AIR MI @ 5.13 NMi/#   COST (-)   137.88#
EQUAL MILE DIFF ACTUAL VS PLAN   +529.52#
6510.32-6267.21 COST TO FLY HEAVY (-) 243.11#
PERF. LOSS COST   286.04 ≃ 286.41#
ACTUAL FUEL   6796.36
THEORETICAL FUEL -6510.32
                   286.04

EXCELLENT AGREEMENT!!!

                                              +39:12:44
75°W TO HOME   +3158.2   -444.07   +3602.3   +927.2 +39.212 Hr
         AVG - 11.325 Knots

         ÷ 4.89 NMi/# = -736.6   FUEL REQUIREMENT VS MIN CURVE

                       190.53   EXTRA REQD VS MIN CURVE PERF SINCE 75W
                                This is the fuel cost to come home
@3825.4# 4.56   rich, fast, high RPM & 2 engine!!!!
@2898.1# 5.22 Avg   4.89 Avg

The Basic Key Data herein is from analysis on the World Flight Specific Range Curve¶ and auxiliary analysis of the 100# climbs and the 286.04# end cost of all losses.

Notice the addition or comparison of fuel weight losses like apples and oranges is invalid until they are converted into equivalent "end of flight" fuel!!!!*

1. For the 6796.36# of fuel used, the minimum curve range of the Voyager is 22539 N.Mi

2. With 21069.8 N. Air Mi on the Voyager World Flight, 1469.2 N.Mi range was lost due to net below curve performance loss, 286.04# of fuel - However

3. 976 N. Mile lost range 190.53# of fuel, was a voluntary decision to come home richer, faster, at higher RPM and untimately on 2 engines once it was seen that we had ample fuel, to protect the cruceal rear engine which ran full throttle from takeoff to Caracas, Venezuela.

4. At 4000# GW just before Trinidad the net loss was 533 N.Mi (120.66# of fuel*)

5. Since early, heavy, on two engines, the plane has flown above the minimum curve and at 7000# GW had gained 153 N.Mi range + 69# fuel* @ 2.21 N.Mi/#Avg, the actual mid flight shortfall had been 533 + 153 = 686 N.Mi (189.6# of fuel*)

6. A climb analysis showed 100# of fuel* would have been used, 376.6 N.Mi range loss @ 3.76 N.Mi/#Avg, in 3 climbs to top Sri Lanka and African weather, 56# to climb, 44# to maintain. (With no climb fuel burn or performance data supplied by the crew, if for example the engine had been run richer than assumed, any error, a percentage of 56#, would not be large since 56 is not itself large.) Note that the 100# climb cost is 73# at flight end.

7. 309.3 N.Mi range loss, 89.6#, was lost in mid flight slow below curve performance.

8. Additionally, 574.4 N.MI, 109# fuel* @ 5.27 N.Mi/# was lost due to a leaking left tip fuel cap.

9. The .0269 correction made to the refined log GW, fuel wt and # leg burn, was not added into the GPH, #/Hr or specific ranges shown, thus graphical solution of losses look slightly low.

10. Total lost range was 2048.88 N.Mi including the 109# fuel lost.

IT IS NECESSARY TO CONVERT MID FLIGHT FUEL BURNED TO EQUIVALENT POUNDS AT THE END TO JUSTIFY FUEL COST VS CURVE REQUIREMENT

	N.Mi	# Fuel		End of Flight Correction	End of Flight Equivalent Fuel
Efficient Heavy Flight	+153 ¶	+ 69	@ 2.21	+69 $\times \dfrac{2.21}{5.13}$	+ 29.725
Weather Climbs	-376.68	-100	@ 3.767	-100 $\times \dfrac{3.76}{5.13}$	- 73.294
Mid Flight Loss Slow	-309.32	- 89.66	@ 3.45	-89.66 $\times \dfrac{3.45}{5.13}$	- 60.2976
Status at 4000# GW Just before Trinidad	-533 ¶	-120.7	@ 4.416	120.66 $\times \dfrac{4.416}{5.13}$	-103.866
4000# GW to 75°W	+ 40	+ 8.88	@ 4.5	+ 8.88 $\times \dfrac{4.5}{5.13}$	+ 7.797
Rich, Fast, Hi RPM, 2 Eng	-976	-200	@ 4.88	-200 $\times \dfrac{4.88}{5.13}$	-190.253
Net Range Loss	1469 ¶	-311.2 ←	A meaningless total, apples & oranges		-286.323 ≅ 286.04
Lost Fuel	579.88	-109	@ 5.32		
Total Lost	2048.88				

*Caution: Fuel cost in #, must be converted to flight end for valid comparison!

```
  2048.88     Range lost  Including 109#
 21069.8      Actual Air Miles
 _____

 23118.68     Calculated range not including 106.14# remaining
   558.29     106.14# x 5.26 N.Mi/#  Sp. Range at end.
 _____

 23676.976    Analysis Range Total

 23693.7      Range for  7011.5#  Fuel Loaded
 _____
```

$\triangle$ 16.7236 N.Mi/5.26 = 3.179# Fuel in 7011.5#

| Error | .000453% | 1/20 of 1% |

At 5.8#/Gal   Error is .548 Gallons in 1171.8

The flight of the Voyager was a great piece of science, but the science is never the whole story. Imagination, creativity, perseverance, people, as well as their good technical work, make things like the Voyager happen. Dick and Jeana and Burt knew how to do the technical job, but the technical rules had been sitting there available all the time, to others, to the whole world. When they, the right people with the dedication and the determination and the creative insight came along, the Voyager happened. It's always a story of both technical accomplishment and people with the know how and guts to proceed.

My real objective after completing the Official Analysis of the Voyager World Flight, a remarkable, indeed magnificent technical achievement, was to tell the story in a way that pilots, people interested in flight, indeed unfamiliar people could understand and appreciate what really happened.

Flight and aerodynamics can be exceptionally complex. Indeed, one of the things that could be seen very clearly in this effort was that no one, not even the best of experts, had their grasp of long range flight completely correct. Essentially everybody had at least one wrong view. Just about everyone's ability to lay out the theory and tell the story amounted to a partial story.

Some of the simple narratives in the beginning of this I believe are quite valuable because they make clear the central truths of flight theory in a way that everyone can understand. A very small percentage of the people involved with flight really understand how the laws of efficient flight work and interrelate. Hopefully, as a result of this presentation, more people will come to have a clearer understanding of the logic of flight.

IMPLIED VOYAGER L/D

BASED ON BREGUET RANGE FORMULA

The classic Breguet Range Formula used by Aeronautical Engineers can be used to approximate the fundamental aerodynamic facts of the Voyager with reasonable accuracy. A review of the engine curves shows that at the average 10,540 ft. density altitude of the flight, a specific fuel consumption of .4#/HP Hr is a reasonable choice for the rear engine.*

The propellers are capable of operating well above the normal 80% efficiency, but considering the rough rear air presented which the John Rooncz design took into consideration and the broad operating range of the Voyager, conservatively using just 80% as a representative average would seem to be a reasonable choice.

The 3184.18# landing weight used below assumes actual air miles[§], a perfect flight up the minimum performance curve from the actual 9694.5# takeoff weight. <u>This means the minimum performance curve implies a 29+ L/D with a .4 SFC, 30+ considering the less efficient front engine</u> and a higher average SFC.

Since the Voyager tests showed it could perform well above the minimum curve, an L/D well above 30, more like 32 or above is appropriate above the minimum curve. The <u>calculated L/D is 36½ with laminar flow, 28 without.</u>

$$R = \frac{Prop\ Eft}{Eng\ Eft} \times \left(\frac{L}{D}\right) \times Ln\left(\frac{W_{TO}}{W_L}\right) \times 375$$

$$\underset{24,246.704}{\overset{AIR\ MI}{}}{}^{§} = \frac{.8}{.4} \times \left(\frac{L}{D}\right) \times Ln\left(\frac{9694.5}{3184.18}\right) \times 375$$

$$\boxed{\frac{L}{D} = 29.037}$$

*30+ Considering the less efficient front engine,
32+ Above the minimum performance curve.

*Having made the point that the very efficient rear engine must operate at an <u>average</u> SFC of .4#Fuel/HP Hr., a number that would only be wishful before the very efficient liquid cooled Continental engine, it's important to note that with the normal air cooled front engine burning a significant part of the fuel, an average SFC of .42 or more and an L/D well above 30 are more proper numbers.

§24,246.704 Statute Air Miles equals 21069.8 Nautical Miles, the actual <u>air</u> mi.

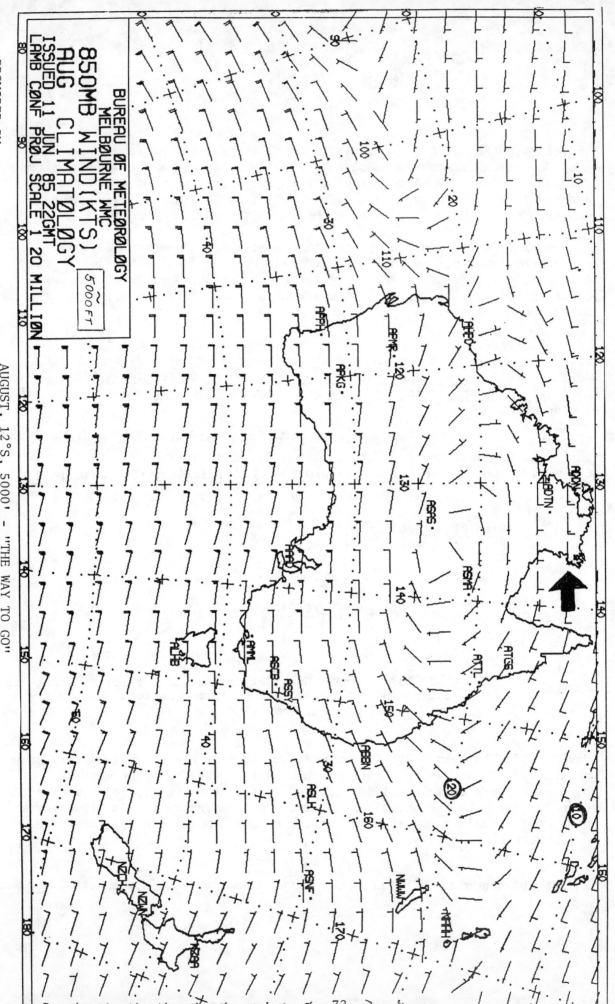

HIGHER ALTITUDE, LATER MONTH, DIFFERENT LATITUDE, ARE ALL LESS FAVORABLE

OPTIMUM LATITUDE VARIES SLIGHTLY VS GEOGRAPHY

BUREAU OF METEOROLOGY
MELBOURNE WMC

850MB WIND(KTS)
AUG CLIMATOLOGY
ISSUED 11 JUN 85 22GMT
LAMB CONF PROJ SCALE 1 20 MILLION

5000 FT

PROVIDED BY
KEITH GORDON
QANTAS

NORRIS'S THIRD LAW: If you fly a World Flight properly the tailwind will almost
always be 8½ knots (±1) (10MPH). The whole earth is our ultimate averager!

AUGUST, 12°S, 5000' - "THE WAY TO GO"

JR NORRIS AUG 86

72